ACKNOWLEDGEMENTS AND MY GRATEFUL THANKS TO Johnny Morris, for writing the preface and for the good friendship, so full of laughter, for well over 35 years; to Mike Curzon, fellow Director; all the staff who have helped over the past 36 years; to James Belsey for helping to edit this book and to Jay Prosser who sorted out and typed off the many tapes which Don and I made when preparing this book together.

THE
ROAD
TO
RODE

BETTY RISDON

The Story of the Tropical Bird
Gardens at Rode

All illustrations by Betty Risdon

Published by
Rode Tropical Bird Gardens, Rode, Nr Bath,
Somerset BA3 6QW

Edited by James Belsey
Designed and set by Patrick Harrison
Printed by Doveton Press
15 Willway Street, Bedminster, Bristol

ISBN 0 9533535 0 8

Contents

PREFACE
BY JOHNNY MORRIS OBE

... *What can you expect?*

An Aunt of mine used to say when Deaths, Births, Tragedies and Scandals visited themselves upon the family: "In this life, dear, you never know what to expect." An innocent enough cliché. But should you be mad enough or blind enough to incorporate into "This Life" a fair chunk of the animal world, then you must know what to expect and that is, generally - the worst. Consider the world of the Race Horse. In many ways it is comparable to a British Rail locomotive. Most impressive to look at, thudding out strength and speed, snorting with conceit and dominance. You cannot suspect that it is plagued with a rash of fatal weaknesses. You don't expect it. You expected a hilarious, prancing winner. You are presented with a flop-eared, limping, steaming failure. You did not expect that. Just look at the faces at a race meeting when a dead cert has just died. Or those alighting from a two-hour-late express train. But, thank goodness, there are those who are forever optimistic and confident and completely undeterred by things that they did not expect. Such a pair were Donald and Betty Risdon.

I first met them at Dudley Zoo in the early 1960s. They were not a bit what I had expected. Donald was an impressive military type, over 6ft tall, with an orderly, well-trimmed moustache. His manner was, to say the least, peremptory. I had with me a BBC producer who was a trifle obscure in his pronouncements. Donald fixed him with a cold blue stare. "I don't understand what you are talking about". He said it at least four times that afternoon. Donald was a bit intimidating at the first meeting. But I soon discovered that he wasn't the barking Colonel he seemed to be. He was a most kind and sensitive man and he was, of course, infatuated by the beauty, grace and colour of birds. And he wanted to concentrate on birds and birds alone. He liked the mammals, reptiles and fish but he was almost possessed by birds. It was this mild mania that set him and Betty off on the Road to Rode.

It was at Rode that I got to know them very well. They were an impressive pair. Betty was only a few inches shorter than Donald and, as a partnership, they tackled the dereliction of the old house and grounds with the energy and optimism of a military campaign. Betty's gardening skills soon began to infiltrate into the bramble tangle that had overtaken the old gardens and her catering organisation soon made the place civilised and attractive. The Crowned Cranes and peacocks were soon strutting as if they owned the place, the Macaws flying fancy free and the light hiss and gentle chuff of the immaculate steam trains now circled the idyllic gardens. This was something that they both had hoped for but very often did not expect. So, perhaps, my Aunt should have amended her dictum to: "In this life, dear, you never know what to expect - so be determined to deal with it and make the best of it."

When you take the road to Rode I'm sure you will agree that Donald and Betty have made a very good best of it.

FOREWORD

I have always found that however much you plan ahead, things all too often don't happen the way you had intended. It happened with this book.

My husband Don and I had often discussed the idea of writing a book telling the story of how we created our all-bird zoo just outside the pretty Somerset village of Rode, not far from Bath.

There was so much to write about . . . how Don's passion for birds began and developed, how we met, those exciting early days at Rode when each day seemed to bring some new challenge, the friends we made, feathered and human, the fun of our many TV and radio appearances and the excitement of our travels overseas to see birds in the wild.

We even had a title - The Road to Rode.

Somehow we never got round to it and the book idea was to put to one side.

But then, in the autumn of 1988, Don began to find problems with his glasses. He was sent to a specialist who diagnosed Macular Degeneration. Neither of us had any idea what this condition was or what it would mean to us and our lives.

Early the following year Don told me that my face was a funny shape. I wasn't surprised at the time and thought he was doing a little of his usual teasing. He wasn't.

Don returned to the specialist for a series of tests. He was told to use a magnifying glass for reading and to give up driving. The only consolation was that he was told he would not go blind.

The shock was immense to a man who had devoted the whole of his life to the care of birds, animals and fish. He had written several definitive books on birds and was very highly regarded by his fellow experts in this specialised field. I thought it might be an idea to cheer him up by

encouraging him to dust off the Rode To Work project and
to start work on the book together. Don turned to me and
said: "How can I, Betty? I can't see." It was a remark he was
to make many times in the following years.

A friend suggested a compromise. Why not get the book off
the ground by tape recording our memories and then tran-
scribing the recordings onto paper? At least it would be a
start. And that is what we did.

All that summer we made recordings, sometimes together,
sometimes separately. I began to sort through our large collection of slides and photographs to find illustrations for
the book.

But later that year Don seemed to lose interest in the project and I found less and less time to type out a manuscript
from all our recordings. Our efforts came to nothing.

Much, much later I decided to make another attempt. I
transcribed our records and put them in my own words,
except for the material about the care of birds. These are
Don's.

Now, at last, the work is complete. It is my tribute to a brilliant 'bird man' who sadly died on April 24 1994 after suffering for five-and-half years.

Betty Risdon, Rode, April 1998.

INTRODUCTION

"Come up and see my animals sometime"

"You must come up and see my animals sometime." That was a pretty original remark, you must agree. It was made to me in a pub in Birmingham.

Early in the day I had been to fetch the actor Esmund Knight who was going to have tea with my mother. As we passed a pub called the Plough and Harrow, Esmund asked if we could pop in there for a drink on the way back. It was one his favourite pubs. By sheer chance, I was going there anyway. It was my birthday and I had arranged to meet some friends for a celebratory drink, so I invited Esmund to join our little party. One of the friends happened to be the advertising agent for Dudley Zoo, a very popular West Midlands attraction.

We were all chatting happily away when a newcomer arrived and he was introduced to me as Donald Risdon, General Manager of Dudley Zoo. Don had just been to a party at ATV's studios in Birmingham and he had clearly had rather a good time. He had even made a date with a girl he had met . . . a damsel of wondrous beauty, he told me later.

We got on very well and, later in the evening, he tapped me on my knee and made his unforgettable remark about seeing his animals. How could I possibly refuse such an offer? The advertising agent was clearly very keen to seize the publicity opportunity of taking a famous actor - rather than me - to Dudley Zoo for a possible photo opportunity for the Press. In the end Esmund and I decided to go together and we had a very interesting time being shown around by Don.

Don rang me not long afterwards to say that they were moving a bear and would I like to come and see how it was done? As things turned out, it took a solid three weeks to move Ivan the bear into his box. I always teased Don that he had spun the time out on purpose so we could get to know each other better.

We saw each other several times in the weeks which followed and gradually I began to learn more about this remarkable new man in my life.

Betty Risdon and the 100,000th visitor

CHAPTER ONE

The Boy with the Back Garden Managerie

Don had been interested in birds and animals since he was a little boy. His father was a civil engineer by profession but he took a keen interest in nature. One of Don's earliest memories was his fascination with his father's aquaria. The family lived in a large house at Kingston-on-Thames which had its own underground billiards room below the basement. The room was cool and dark and ideal for keeping fish tanks in the days before electric aeration and filtration of sea water.

The large local department store Bentalls had its own zoological section which sold many kinds of marine specimens and Don's father was a regular customer. Don vividly recalled the moment when he noticed that a crab in one of his father's tanks had just changed its shell. His father explained what had happened and why and the small boy was struck by the old shell lying discarded in a corner while, under a nearby rock, the now much enlarged crab sheltered.

Don spent hours watching elegant prawns, flower-like sea anemones and tropical fish like Blenny. Looking after tanks and their inhabitants was hard work in those days. Don's father would bring home casks of sea water from Bentalls and cool it in the billiards room before siphoning it into the tanks. Then, by using a system of glass tubes, air was bubbled into the water to keep it aerated, an essential for the marine creatures. Don made good use of this Heath Robinson-like equipment when he began to keep fish of his own.

The garden had a pond with a fountain which was home to frogs, newts and toads which spawned in the water. Within weeks of spawning, the garden would be full of half-grown toads and frogs which Don collected and kept in enclosures and tanks. This was his initiation into the art of caring for small creatures.

Richmond Park was not far away and its ponds and pools were a constant attraction. Don's father regularly gave up his Saturday afternoons to take his son pond-dipping. They would set off with long-handled nets and glass jars to collect specimens like dragonfly larvae, water boatmen, water scorpions and water beetles of various kinds. These would be triumphantly borne home so that Don could study them.

World War I brought a complete change of scene. The Germans began a bombing campaign against London. It was very limited by World War II blitz standards, but it was enough to make many families decide to leave the capital, including Don's. They chose Bournemouth.

The move brought Don into contact with a completely new environment. He explored the local lanes catching moles and butterflies and he also discovered the lizards and slow worms which were common in the area.

He and his younger brothers and sister were taken by their nanny for walks to chines - valleys which run down to the sea - like Alum Chine and Branksome Chine and allowed to explore while their nanny rested on a bench. It was on one of these outings that he caught his first slow worm. He carried it back triumphantly to his startled nanny who showed it to a man who was walking by.

"Get away with it" the man said urgently. "It's a snake. It will bite you and kill you".

Don knew even then that slow worms were harmless, legless lizards and he burst into tears as he pleaded to be allowed to keep his prize. Carelessly, he put it down on the grass and when he turned back it had burrowed into the turf and vanished.

Don's new surroundings on the warm South Coast were very different to Kingston-on-Thames. He would search beneath the bracken and brambles along the banks of chines and get tantalising glimpses of all kinds of lizards. He told me that he was sure they were Wall Lizards and not the ordinary common brown lizards. They sparkled with green and silver and were very beautiful. He did his level best to catch one but never succeeded - they were always far too fast and agile.

When the war ended, the family returned to Kingston-on-Thames and a new home and Don and his father resumed their pond-dipping expeditions to Richmond Park. Here they found all-too-moving reminders of the war when they saw Service invalids from the Royal Star and Garter Home on Richmond Hill. Many of these veterans of the trenches, dressed in blue uniforms to denote their status as war-wounded, used the park for recreation and their injuries and lost limbs made a lasting impression on Don.

Despite their disabilities, the veterans were lively and full of mischief. They would tease Don by opening his boxes of collected butterflies and letting them loose. It was probably their way of showing off to his nanny, a young and attractive girl. She would sit on a park bench and the soldiers would flock to her to pay her compliments . . . those trips to Richmond Park became a daily event. While she and the veterans flirted, the boy happily roamed around looking for wildlife.

The new home had nowhere suitable for Don's father's aquaria but it did have a veranda which Don was allowed to use for his collection of creatures in rows of jars and bowls.

One summer was particularly dry. Don brought a frog back which took to sleeping in the top of a mud-filled glass jar. It made a little hollow for itself in the mud and stayed there for weeks while the drought lasted, clambering out from time to time to swim in a bowl where he kept his pond specimens. The family told him that on one occasion they had seen the frog return from the garden, climb the leg of the table on which the bowls and jars stood, hop into its mud shelter and then immerse itself in the bowl of water. Don could never quite believe the story but his family swore it was true. Apparently the frog would vanish into the garden by night and then return to where it knew it could find water. When the rains eventually came that autumn, the frog disappeared forever.

In 1920 Don's father decided to move the family to the seaside. He felt it would be a healthier place for growing children - and a safer place. He was one of those who felt convinced that there would be another war with Germany because of Germany's lasting resentment of the Treaty of Versailles.

"You can depend on it - we shall be at war with Germany before long" he would say.

The family moved to Hove, near Brighton. It was a very smart place and considered itself very much superior to the neighbouring seaside resort. The house had a small garden which proved a paradise for the young naturalist. It was here he really began to build up a menagerie. His father decided to encourage his son's interest by building a vivarium in one corner of the garden. A concrete trough was created to hold water and a bank built inside with clinkers covered in soil.

When Don's parents went shopping together in London, Don's mother made a point of visiting the zoological department of the famous Knightbridge store Harrods to choose some new creature for her son's collection. She brought back white mice, guinea pigs, Green Lizards and Green Tree frogs and, once, a pair of green budgerigars for one of Don's sisters. She also brought mealworms for feed.

Don soon had a collection of black and yellow Salamanders, Green Lizards, grass snakes and other creatures. As he became more and more interested in birds, the vivarium was extended into an aviary and he began to stock his little zoo with small aviary birds which thrived and did not interfere with the reptiles below.

Don's knowledge of the care of animals and birds increased all the time. He also began to learn more about breeding birds. His mother suggested that he should start with canaries and he was presented with a pair one birthday and allowed the use of an attic room in the house.

He bought a book on canaries and was thrilled when nature took its course. The pair built a nest and the cock started to feed the hen. One day, to his utter delight, Don found four eggs in the nest. He never forgot that moment for the rest of his life.

The little hen hatched the chicks, to Don's great excitement. He watched the tiny creatures with their open beaks waving about in the nest but, sadly, he grew far too enthusiastic and overfed the parents with a biscuit food mixed with hard boiled eggs. The cock and hen sickened and passed on the infection to the chicks and the whole family died.

The infection was known as 'Sweating Hen' in those days. People then believed, quite wrongly, that birds sweated like humans and that the sweat from their body drowned the chicks. In fact birds do not have sweat glands like ours. The problem, he was to discover, was bowel trouble caused by wrong diet or overfeeding.

Don turned to a canary fancier for advice. He was told: "Don't give them great dollops of food all at once. Feed them little and often." Now he knew how to successfully breed canaries.

Later he began to take an interest in pigeons. It was the start of his life-long love for these birds. Some teenaged boys who lived next door invited him to see their pair of homing pigeons. The pigeons were kept in a hutch and the hen was already sitting on eggs in a nest box at the back.

Don fell for them at once. His neighbours built a loft in their garden and the birds continued to breed successfully. When they grew older, the young men lost interest in their pets and asked Don if he wanted them. Don had several friends at school who were also keen on pigeons and they went in a group to the house to choose them. They each ended up with two pairs apiece.

With the birds safely settled in their home, Don was keen to attempt some homing flights. It quickly turned into a craze. He and his friends walked or cycled five miles, released the birds and, sure enough, the pigeons were back home in a matter of minutes while their owners made the far slower journey on foot. Don began taking birds to school so he could send messages to his brothers and sister at home.

The family didn't own a motor car, however Don's father did commute daily to his job in London by train and Don persuaded his father to help him experiment with long-distance flights. His father would step off the train at one of the stations en route and release the birds. They always returned safely to their loft.

Although Don loved his pigeons, it did not distract him from a growing interest in other creatures. A cousin who bred fancy white rats gave him a pair of females which were always getting loose in his attic room and running around the top floor. Don's father detested rats because he knew what a menace they could be. Each time they escaped, he was terrified they would breed with local wild rats and that the family home would become overrun with piebald, half-wild rodents.

A school friend gave Don a white English rabbit with the usual black markings. He loved his rabbit and named it Spot after the black spot on its nose. Spot lived with him for years.

If you make a real pet of a rabbit and give it a chance to develop properly instead of just keeping in a hutch, it quickly shows its natural intelligence. Spot was given the run of the garden every day. The garden was small and surrounded by high brick walls, which made it perfectly safe. Spot was soon playing games with the children and if Don ran away from his rabbit, Spot would chase after him and run in tight circles around his legs to stop him moving away, as if to say: "I've caught you - now you can't escape!"

Don's parents proved remarkably tolerant. Whatever their private misgivings, they even allowed Don to buy a monkey. I suppose most boys would dearly love to have a monkey.

Don had read a great deal about monkeys and what marvellous companions they make. One day he visited a pet shop and saw a young Rhesus monkey which had been hand-reared and was completely tame and amenable. He longed to buy it and his parents agreed that he could.

The pet shop owner said it was a South American capuchin monkey but Don knew better. When he visited London Zoo he compared his with the zoo's very fine collection of Rhesus monkeys. When he saw the huge males, he was horrified by their size and the thought that this is what his little pet would become.

He called the monkey Jacko. Monkeys do make good pets, but only for the first three years before they mature. In the wild, this is the moment they take charge of their own group and move away to new territory to become independent.

But for that first couple of years Jacko was a great companion. Don quickly learnt how much Rhesus monkeys love water. When he was still very small, Jacko made a beeline for a water butt in the garden and happily splashed around. Don asked friends if they would like to watch Jacko having a bath. He filled up an old-fashioned tin bath with warm water and Jacko was in his element, diving in, swimming underwater, jumping out and then shaking himself over the boys.

Don took him into the nearby woods and released him. Jacko shinned up and down the trees and jumped in and out of any ponds he could find.

Don built an enclosure for Jacko in the garden, but he proved an expert at escaping through weak parts of the wire. Once out, Jacko would dive into the house. One day, when Don had been away in London, he arrived home to find an ominous silence and scenes of devastation when he walked through the front door.

Jacko came shooting down the banisters and landed in his arms. Don heard tearful voices behind a locked door. It was his mother and youngest brother, both armed with golf clubs. They had heard Jacko come into the house and were too frightened to emerge from the room. They had to listen impotently to the sounds of destruction upstairs as Jacko banged and crashed about. Don's father's dinner jacket, which had been laid out on the bed in preparation for some function that evening, lay in a heap at the bottom of the stairs.

Another time, Jacko escaped and climbed into a neighbour's apple tree which was the pride and joy of its owner, an elderly lady. He leapt about in the branches, hurling down a hail of apples. Then he spotted the lady's cat. The cat ran up the tree but every time it tried to attack Jacko, he jumped clean over its back and began pulling its tail.

The poor neighbour who had shut all the doors was leaning out of her bedroom window shouting: "Oh! My cat! Oh! My apples!" Don rescued Jacko - by now he was the only person Jacko would come to.

Everyone had become terrified of Jacko and he had to go. Don went on to keep other monkeys including a capuchin, a second rhesus and a green monkey but by now he had learnt his lesson and built a proper cage of much stronger material.

All this must seem strange to modern readers. Certainly no boy would be allowed to keep a monkey as a pet. Today monkeys are classified as dangerous animals and a special licence has to be obtained before the authorities allow you to keep one. But when Don was young you could buy any animals or bird you wanted provided you had the money.

Don's collection reached menagerie proportions and many friends and neighbours came to see his birds, parrots and animals.

One fateful day a visitor remarked: "This is just like a zoo . . . you really ought to charge 2d a time to see it."

Don did some mental arithmetic about how many pennies he could make and it set him thinking. That casual remark sowed the seed of an ambition - to run a proper zoo of his own one day.

These carefree days came to a close. After he left school, Don trained as a chartered secretary and before long he had his first full-time job in an office in the heart of the city of London, commuting daily from Hove. He hated it. Clerks had to wear formal clothes and Don detested wearing suits, starched collars, ties and a hat. In hot weather he found the heat and confined concrete streets of the City absolute purgatory.

Don had never wanted a career like this - he longed for a life in the open air. He stuck at the job for 18 months, becoming more and more miserable. He became so depressed that he felt that if this was to be his future, he might as well not be alive.

Don began writing round to all sorts of places, including farms, in search of an outdoor job, any outdoor job, where he could use his skills with birds and animals. One was the Keston Foreign Bird Farm in Kent, thought to be the only farm of its kind in the world at the time. The farm imported and bred a wide variety of exotic birds including Australian parakeets. To Don's amazement, he received a reply saying that although they didn't have a job at Keston, they were, however, considering opening a shop in London.

This was not exactly what Don had in mind. It would

mean working in London but at least it would be a job in which he could take a real interest.

He went for an interview at Keston with its owners, Alec Brooksbank and Geoffrey Boosey. It all went extremely well and the two men could clearly see Don's keen interest in birdkeeping. Don was shown around and was enchanted. The farm was near Biggin Hill and looked across the valley to Keston, then a pleasantly quiet country village. The farm had a lovely collection of birds, all beautifully housed in movable aviaries. Everything about the place was perfect.

Don accepted their offer of the London job, hoping it would eventually lead to a job at Keston itself - which it did. The shop was not a great success but at least Don was dealing with birds and making a living from his passion by selling to customers who were not prepared to make the journey down to Keston.

When the shop was finally closed, Don was invited to become manager of the bird farm. He was thrilled. It meant a complete change of life. He had to leave home and live in digs, a new experience, and settle into an entirely new routine. Now, at last, he was doing what he really wanted, caring for rare and fascinating birds. All his other interests in monkeys, squirrels, guinea pigs and the rest faded into the background.

Keston had become quite famous by then and had many breeding "firsts". Don helped breed the first Blue Fronted Amazon Parrots in Britain. In those days that was a considerable achievement, although today it is not unusual thanks to big improvements in diet.

The cock and hen had been kept together by their owner Lord Tavistock, later the Duke of Bedford. The couple had become quite savage and he could not get breeding results. Chicks would hatch but the parents persistently let them die.

Geoffrey Boosey, a very intelligent man, thought about this problem and decided that the birds needed more than seed to rear their young. The birds were moved to Keston and he experimented with various foods including boiled rice pudding and cooked fish - most un-parrot-like mixtures! But it worked. The food was soft and the birds could regurgitate it to feed their young. In that first season they managed to rear five beautiful chicks. The parents became even more savage with the young so they were placed in separate shelters.

Other British "firsts" included Roseate Cockatoos and Australian Parakeets, all thanks to successful experiments with different diets to help the parents feed their young.

But Don's idyllic existence was soon to be shattered. War clouds were gathering and on September 3, 1939 Britain declared war on Germany. Don decided to volunteer. If you volunteered rather than waiting for your call-up, you could choose which Service you wished to join. Don was determined to join the Royal Air Force. He wanted to become a pilot because he felt if he was to be killed, he would rather it was over quickly. He had grown up with tales of the Western Front in World War I and had no intention of experiencing a hellish life in the trenches.

He wrote to the Air Ministry and was selected for the equipment branch. His job at Keston came to an end - they could no longer afford him because of the war - so Don joined the Auxiliary Fire Service to keep himself occupied until his call-up. It was voluntary service and he learnt drill and how to run up and down ladders carrying bodies from upstairs rooms and how to put out fires.

Don served for six years with the RAF. He never became a pilot after all, but served as a technical officer in several countries. These overseas postings gave him plenty of opportunities to study very different wildlife to what he had known so well in England when the demands of war were not occupying his time.

Don spoke very little of his war-time experiences. It was an interval - even if a very interesting interval - between his real career of raising, breeding and caring for birds and animals and it was a very happy Don who returned to Keston after six years, this time as a Director of the Company.

Don stayed at Keston for a further four years and by the end of the 1940s he had become a leading authority on the care of exotic birds.

In 1949 he was offered a new post, as General Manager of Dudley Zoo.Don accepted the new, far more responsible and challenging job but even before he had arrived at Dudley he had formed an ambition to run a tropical bird zoo of his own.

Dudley would provide him with the perfect opportunity to improve the management skills he would need if he was to start his own business.

CHAPTER TWO

Learning My Lessons

I grew up in a very different world to Don's and with very different hobbies and interests. I am sure this is why we went on to become such an effective team. We complemented each other perfectly when it came to the exciting challenge of setting up our bird gardens at Rode. My family came from a mining family in North Yorkshire. They lived in Saltburn and had a summer house in Hinderwell to which they eventually moved.

My grandfather, William Walker, held a position of some authority in the mining world. I never knew him, but I did hear one story about his talents. He was a very gifted inventor who, sadly, never patented any of his innovations. One day he returned home in a very distressed state. A cage at one the collieries had broken and fallen and all the miners had been killed. He was determined that nothing like this should ever happen again.

As he wrestled with the technical problem of preventing cages crashing down mine shafts he began to play with a pair of my grandmother's scissors. As he opened and shut them, an inspiration struck him. The result was the invention of the detaching-hook, based on the way a pair of scissors springs open. He earned the nick-name "Hookey Walker" and I treasure the little scale model which has been in the family since his days.

He had three sons and four daughters. My father was the youngest and his eldest brother was 26 years older than him. The two brothers both became Chief Inspectors of Mines in their own right. My father worked as manager of a pit outside Birmingham and his family lived in

Birmingham, where I was born.

My grandfather on my mother's side was a brilliant doctor from Inverness who had taken his practice to Redcar, where I often visited him. He died just before the advent of the National Health Service. He was a very caring man who went out day and night to look after his patients. All too often, he forgot - or perhaps never bothered - to send a bill. I think both families would have been amazed by today's mining and medical fields.

When I was about seven, my parents took me on a cruise to, of all places, Norway. I do recall how beautiful the country was but little did I know that it was to be the first of many overseas visits.

Not long after our Norwegian holiday, my father was invited to open some coal mines in New South Wales in Australia. We travelled out by ship. The sea was very rough at times and the voyage seemed to last for months.

When we arrived in Australia, we stayed in a small town called Cessnock, nearly 200 miles inland from Sydney. The hotel where we stayed was rather shabby and noisy and when my parents went down for their evening meal, I was left behind under a mosquito net. Unfortunately the net was riddled with holes. The noise downstairs and outside the hotel was deafening. A man was loudly playing "Moonlight and Roses" on a guitar and the sound of traffic was incessant. The next morning my parents discovered to their horror that mosquitoes had flown into my net and that I was a mass of bites. I looked as if I had measles.

We moved out into the bush about six miles from town. Our house overlooked the pit and a dam with hills in the background. The garden was full of fruit and vegetables with masses of gum trees beyond. My parents installed a septic tank so that we could have an indoors loo instead of a privy at the bottom of the garden. They also built a tennis court where we held lots of tennis parties. Our visitors were much more impressed by the indoor loo than our tennis court!

We were surrounded by wild countryside. Those six miles to Cessnock to get supplies took at least an hour because the roads were so bad. At times we were completely cut off by floods. We suffered droughts, dust storms and, of course, bush fires. The last really big bush fire happened just before we returned to England.

I was sent to boarding school in Sydney, overlooking the harbour and the bridge. Out of self-defence, I rapidly acquired a good Australian accent - much to my parents' horror. I was the only "Pommy" in the school and the other girls really bullied me because of my English voice. If you can't beat them, join them . . . and I did just that with my Australian twang. Whenever I came home on holiday, my parents strove to change my accent back again.

The school took us on outings all over the place. We swam in an open pool in Sydney harbour which must have been pretty filthy. The moment you could manage to stay afloat, you were forced to jump into the deep end. The first time, I came up spluttering and I can still remember those bubbles. But it worked and I thoroughly enjoyed swimming and soon got to know all the best surfing beaches in the area.

I grew to love the sporty, outdoors life of Australia and have a very soft spot for both the country and its people. The sand in my shoes from Australia is still with me and, I'm sure, always will be.

We spent five years in Australia until my father's contract ended. We sailed back to England and returned to Birmingham, where I was sent to school.

My father was waiting for a job as an Inspector of Mines so, to fill in the time before his new appointment, he took an agency job selling colliery explosives. My mother helped him in his work and travelled around with him from mine to mine. But my father's health deteriorated and one day at school I was sent for by the headmistress. She broke the news that my father was dead. He was only 40 years old.

My mother had to become the bread-winner. She wrote to the explosives firm and offered to continue my father's work. Perhaps out of pity they agreed that she could but they probably thought she wouldn't last more than a couple of months. In fact she went on to become their very best agent! She added miners' helmets to her sales list and appeared on the pit banks. No woman had ever done work like this before. During the school holidays I would trail around the pits with her.

At school, I was pretty hopeless at most things except for cricket, tennis, history and geography. I had to abandon my attempts to learn Latin, something I very much regret. It would have been invaluable when I began to work in the bird world with its Latin classifications of birds and bird families. Maths was not my strong point either, but you can soon teach yourself how to count money and keep accounts, as I was to discover much later.

After I left school I did a shorthand and typing course with modest success. Luckily these days I have a computer with a "delete" key to clear up my typing mistakes!

My spare time was spent ice-skating and riding my horse called Moonlight. Later, I became very keen on rock-climbing and ski-ing. I climbed in North Wales with the Midland Mountaineers and did some climbing in the Swiss Alps.

My mother sent me to the Gloucester Domestic Science College to learn industrial catering. The college was run on very strict lines compared to these more easy-going days, but what I learnt there was to later prove invaluable.

The college philosophy was simple: If something you did was good, you could always do better if you worked even harder. This reminded me so much of my mother. She was always saying: "I can do it, so why can't you?" This was the spirit which had helped her when my father died and it was a trait she had inherited from her father.

I think I must have inherited that attitude too because

instead of bumping along at the bottom of the class as I had done during my school days, I now found myself among the high-fliers. Now at last, thanks to my success in my catering studies, I had found something I was good at. I could cook. After leaving college I had a variety of jobs with tea rooms before I joined the war-time W.A.A.F.

I had to undergo the usual square-bashing at both Blackpool and Southport before I was sent to Loughborough Officers' training course and then to London for a Catering Officers' training course at the time of the VI 'doodlebug' attacks.

I was eventually sent to Skegness where I met and married my first husband. After 18 months, I left the W.A.A.F. to have my elder son. I went to stay with my mother who had moved to the country because of the fear of bombing raids on Birmingham. Another son followed but the marriage, like so many war-time marriages, did not last. My mother and I and my two young sons moved back to her flat in Birmingham and soon the war was over.

Years later I was back in the Skegness area, although only briefly. I was sent to Mablethorpe to take charge of food for the W.V.S. at the time of the dreadful east coast floods in 1953.

Back in Birmingham I joined the local tennis club at Priory in Edgbaston and made many good friends as I began to rebuild my life. I volunteered to help the club with its catering and was soon doing the catering for events ranging from small squash suppers to large tennis tournaments.

I helped my mother with her work and also joined the W.V.S. so that I could make better use of my catering skills. I became part of its Food Flying Squad, a team of experienced caterers who could be called in at times of emergency. We handled everything from training exercises to disasters like the east coast floods.

One was a big train crash when the driver had taken a

corner too fast at Sutton Coldfield. We worked throughout the night on the platform, feeding the rescue team. The damage was horrendous and it amazed me that despite so much wreckage, many passengers had survived the smash.

A year later I unfortunately hurt my back. There was one wonderful consolation. My younger son Andrew's god-mother, an American who lived in Florida for part of the year, invited me to join her while I recuperated. I left the boys in the care of their nanny and my mother and the trip did me the world of good. My back recovered well and I thoroughly enjoyed being taken all over Florida.

It was while I was in Florida that I visited Parrot Jungle, just outside Miami. I enjoyed it as much as anything I had seen in the United States. I was fascinated by all the bright-ly coloured macaws. Little did I know that not so many years later I would be hand-rearing our own! I returned from Florida feeling so much better.

All these experiences were to stand me in very good stead for what lay ahead. I understood catering, I knew how to cope with emergencies and, above all, I could be flexible and inventive when it was needed.

And then I met Don . . .

CHAPTER THREE

A Zoo Director's Wife

Don and I had seen each other several times and I had visited him at Dudley Zoo when one day he surprised me by asking if I would like to man a turnstile at the zoo over Bank Holiday Monday. We all went to help - my mother and my sons Keith and Andrew.

Keith was put in charge of an ice cream kiosk, Andrew found himself looking after the dodgem cars. My mother and I ran the turnstile. We were assured it was a quiet one. We had 11,000 people through that so-called 'quiet' turnstile that day! It was operated by a foot-pedal and I could hardly walk for days afterwards.

That day was a real eye-opener for me. One woman tried to push her pram through the turnstile, another continued knitting as she passed me and her wool became wound in the mechanism. One man came up to me not long after he had paid his entrance fee and demanded his money back. He said that all he could see was people, not animals. That day the Zoo had no less than 56,000 visitors, so perhaps he had a point.

I could hardly believe the size of the crowds. Just before the Zoo closed at the end of the day, one of the keepers asked if I would like a bottle of beer. It tasted like nectar after all my exertions.

This was my first real insight into the working life of a zoo and I found it fascinating to see life from the other side of the counter. Of course I already knew Dudley Zoo pretty well - I had often taken the boys there - but now it took on an entirely different meaning as I began to see the animals from close quarters.

A few weeks later Don invited me to accompany him to an Avicultural Society meeting at Chester Zoo. I felt I was going into the lion's den with some of the world's leading zoo and bird experts. I was very nervous and wore my smartest hat and gloves as I prepared to meet so many of Don's knowledgeable friends and colleagues from this very different world.

Once at Chester, I was quickly put at ease by the charming George "Motty" Motterhead, Director of Chester Zoo. He was extremely kind to me and we were to become great friends. He very sweetly ignored my ignorance, especially the moment when he asked me what I thought of his lechwe. I hadn't a clue what he was talking about! Was it a bird, a fish, something to eat? In fact it was a type of antelope. At the end of my guided tour I was introduced to everyone at the meeting and they were all delightful. I could only assume I had not dropped too many clangers. I had gingerly survived my first test and, I like to think, I had won a quiet nod of approval from Don.

Don and I began to see more and more of each other and after we became engaged, I thought it was time that Don learnt something about my life. It was very different to his world of animals and birds. Mine revolved around sport and physical activity - tennis, squash, skating, ski-ing and climbing.

I rather daringly invited him to join my sons and I at the Midland Mountaineering Hut for a scramble up Tryfan.

My son Keith is an agile climber and he and I had no difficulties with the ascent. We got Don and Andrew half way up and then they announced: "We're not going up and we're not going down". Oh dear!

Keith and I managed to get the pair of them to the top with the help of a rope but Don hardly on speaking terms with me, particularly after he saw a dog managing the climb easily. Don muttered darkly that the dog had the considerable advantage of four feet as opposed to his two and that he might have done better to bring one of the elephants from Dudley Zoo. I was afraid that our wedding, already planned, would be off, but Don forgave me.

As the reader will have gathered by now, Don wasn't a sporty type at all. At school cricket games I'm sure he must have spent as much time as he could fielding in the deep so that he could look for insects rather than retrieving the ball.

On another occasion, I took Don to watch the tennis at Priory Tennis Club in Birmingham during a pre-Wimbledon tournament. I helped with the catering at the club and had made friends with some of the top international stars who came to Edgbaston each year to warm up before the big event.

Carmen Lampe, the USA's No 8-ranked player, was staying with me and kindly asked Don and I to be her guests for the day with seats on the centre court, lunch and then a cocktail party. She gave us lunch in the players' restaurant overlooking the courts. Don gazed down, saw the umpires and asked in a loud voice who were "those able-bodied gentlemen sitting around the edge doing nothing." Something very similar happened when I took him to a Test match at Edgbaston County Cricket Ground. That was it. I never risked taking Don to a sporting event again.

After our wedding I quickly had to adapt to my new role as a zoo manager's wife. We lived in a flat above the office at Dudley Zoo and the telephone was switched through to our apartment out of office hours. Very late one night I took a phone call from a drunken man who tried to sell me a water buffalo. I hardly felt like buying a water buffalo well after midnight!

On another occasion I was in the middle of preparing an elaborate dinner for my new brother-in-law Jack, who was a director of the famous Birmingham car components firm Joseph Lucas. Jack liked the good things in life, so everything had to be perfect. Our front door bell rang and I was just about to fling my arms around Jack when I saw a lady in a fur coat carrying a brown paper parcel.

"Please can you help me" she begged as she held out the parcel. "Could you carry out a post-mortem on my Axolotl?" I hadn't a clue what she was talking about. Don was standing in the background.

I managed to blurt out: "I'm so sorry, but the zoo office is closed for the night. Could you come back tomorrow morning?" I could feel Don's silent laughter almost rocking the flat.

She pleaded: "I've been to the fishmonger and he can't help me . . ."

Ah, I thought, it swims . . "but if I leave it until tomorrow it will go bad."

Don, when he had recovered from his mirth, took pity on me and came to my rescue. I then learnt that the mysterious Axolotl was, in fact, a type of salamander. I learnt to take these sort of incidents in my stride.

Not long afterwards a television crew arrived to record a programme in one of the large storage buildings at the zoo with the well-known TV presenters Peter West, Maxwell Knight and Stanley Dangerfield. All three celebrities came back to the flat after the recording and I served them tea and tried not to be too over-awed.

Don and I then had to dash off to London where Don was helping to judge at the National Cage Birds Show. The M1 had just opened and it was a great thrill to speed along the brand new six-lane highway, but a thrill tempered by my feeling increasingly unwell.

When we returned two days later I felt even worse and disgraced myself by forcing Don to make an unscheduled halt on the motorway's hard shoulder so I could be sick. Don took me to my mother's flat and jaundice was diagnosed. I shall never forget Don's face when he brought me a bunch of daffodils - he could not decide which was more yellow, the flowers or my face.

Once I was back on my feet, I began to learn to give as good as I got. I was manning the switchboard at the zoo on April 1, April Fool's Day.

Various jokers had left phony messages for their friends and business colleagues and given the zoo's phone number.

I knew what to expect and I was well prepared.

My first caller was anxious to speak to a Mr C. Lion.

"I'm so sorry but Mr Lion is in his bath."

"Oh dear, how long will he be?"

"All day, I'm afraid."

"Is there another Mr C. Lion? I was told he needed to speak to me urgently."

"Yes Sir, there are more but I'm afraid they are on their way from America at the moment."

At last the penny dropped and the caller realised he had been April Fool'ed.

I had a lot of fun that day. Mr Fox was out - he was being hunted. Mr Mac Aw was up a tree. Miss L.E. Fant was busy packing her trunk to go away and as for Miss G. Raffe, well, she had a sore throat. The calls never stopped but while some people took it in good part when they realised their friends had fooled them, others were furious. One caller, from the Planning Office, was particularly upset at being caught out and slammed the phone down on me.

One day, Don said "We need some sea lions collected - they're due in from Los Angeles. Would you and Andrew like to go to London Airport to fetch them in the zoo van?"

I set off with Andrew, my younger son. London Airport was a very different place in 1960 to the Heathrow of today. International air travel was for the few in those pre-jumbo jet and pre-package holiday times. Once at the airport, we made our way to the Import Embargo section. The man behind the counter was in a terrible state because some important television film had been mislaid.

When our turn came at last, he snapped: "And what do you want?"

"Four sea lions, actually."

He clapped his hands to his head in despair and said: "My God - that's all I need."

It took a little time to discover that the sea lions were still in Copenhagen and were not due to arrive until the next day. I rang Don to tell him what had happened and Andrew and I drove back to Dudley.

We returned to the airport the next day and were directed to the RSPCA centre. The sea lions had turned up at last and had been well cared for and given a good feed of fish. It was a particularly hot day and our van was very small. One of the cages had to be wedged almost between the two of us and its occupant gave horribly fishy burps all the way back to Dudley.

On another animal pick-up, I was sent to Liverpool Docks. The security police wanted to know what I was collecting. They were astonished when I replied: "A kangaroo." The kangaroo was fine. It had been looked after by the ship's butcher who carried it down the gangway.

Accomplishing these missions was a novel experience for me and great fun. I soon became quite adept at transporting livestock. It was to stand me in good stead for the future.

CHAPTER FOUR

To The Manor Borne

Don had made no secret of the fact that his greatest ambition was for us to open a zoo of our own specialising in exotic birds. His passion for birds was as strong as ever, although he loved looking after the mammals and reptiles at Dudley Zoo.

Don's expertise was widely acknowledged by bird-fanciers and his books including "An Introduction to Bird Keeping" and "Foreign Birds For Beginners" were already standard works.

Don felt that in conventional zoos his beloved birds were all too often overshadowed by the larger animals like elephants, giraffes and the big cats. Birds deserved to be presented in a more sympathetic way, he believed. There was the added charm that birds were not dangerous and presented no threat to visitors.

Whenever we discussed his plan, I always supported him. I suppose I knew in my heart of hearts that my role would almost certainly be that of a dogsbody and a galley slave, but I bravely told him: "Yes, let's have a go on our own".

So we began to search for a possible site. We kept our plans secret because Don didn't want his colleagues to know that he was thinking of leaving. We began to taking magazines like The Field and Country Life to scan the property pages. One of the first places we thought might be suitable was in the south of England. We didn't know at the time that the property was owned by Lord Montague and the Beaulieu Estate. We decided to take a look.

After we had viewed the place, the estate's Land Agent asked us what we felt. We let him into the secret of what we

The ruined mansion of Witley Court.

were planning and told him we were concerned at the lack of potential car parking space. This was at a time when many stately home owners were trying to dream up ways of increasing visitor numbers to help pay the upkeep of their ancestral homes. Lord Montague already ran a successful jazz festival and classic car museum, and our bird zoo might fit in with his plans, the Land Agent told us.

Not long afterwards, his Lordship asked us to lunch. Again, the social dilemma. Should I wear a smart hat and gloves? The meeting was friendly and informal but we hit a snag when he told us that there was nowhere for us to live on the estate. Perhaps to help us overcome our very clear disappointment, he rushed out and picked a bunch of daffodils which he presented to me.

Inevitably, the news that Don was thinking of leaving Dudley Zoo leaked out and our plans were made public. Almost immediately we had a phone call from a man who said that he knew the perfect place and that we should see it at once. It was called Witley Court. When I asked him if it was large, he told me that it was even bigger than Buckingham Palace and had lakes. Frankly, we had been looking for something a little more modest. Don grinned when he heard my news but agreed to take a look. I had never heard of Witley Court. I rather think Don had.

We drove through the Worcestershire countryside to the village of Great Witley and, to my horror, Witley Court, the bigger-than-Buckingham-Palace property was also roofless and ruined and had been largely destroyed by fire in the 1930s. The grounds were overgrown and melancholy and the once-immaculate main lawn was dominated by an enormous fountain with a vast statue of Perseus.

I turned to the owner and asked: "Where on earth do we live?" He said he had just the place. We drove down to a nearby house which seemed perfectly acceptable until I asked about the water supply. He dived into the kitchen and began moving a large handle. It was a water pump!

The melancholy grounds of Witley Court dominated by the enormous fountain and statue of Perseus.

Part of the formal gardens at Rode Manor in Edwardian times.

"Very good for your health and your stomach muscles" he said cheerfully as he pumped away.

Don saw the panic on my face as I had visions of having to fill a tin bath for him. We made our excuses and left.

I was fascinated to read the other day how that sad, neglected ruin with its overgrown grounds has been transformed. The grounds have been restored and the ruin made safe as a romantic reminder of the heyday of grand country mansion living.

Our search was now in earnest and we scoured the country. Before long we had been offered everything from a castle to a rubbish dump. The castle was Lympne Castle. The rubbish dump was in the Lake District.

We were sent to places with incredible-sounding names . . . Sludge Hall, Pyes Farm, Fathoms of Beer and the Valley of the Bird. We were even offered a property which was being run as a Nudist Colony. It was in Sussex and miles from anywhere. We had to keep getting out of the car to open and shut gates which had been set up to protect the privacy of the naturists.

We gave Bleak House a miss but Hackerty Way in Devon was lovely though, sadly, quite unsuitable for our plans. One particular house was advertised as having a wonderful view from the bath!

We visited a place owned by a pair who were the spitting image of Tweedledum and Tweedledee. There was a huge room with a solitary old-fashioned bath. Whenever we tried to find out the price of the property, the two funny little men rushed into the bushes to confer before returning with their answer. This happened several times. I recall another place owned by a gentleman with bright yellow socks and a four-poster bed.

At Kenilworth in Warwickshire we were shown a rather dilapidated building in the middle of a flat field. The kitchen table was alive with cats and we refused the proffered cups of tea. We asked to see the bedrooms and were

Rode Manor in its heyday.

The grounds at Rode as we first saw them.

taken up a very narrow, winding staircase to the main bedroom. To reach the second bedroom we had to go down on
our hands and knees and crawl through a low archway -
very difficult for us since we were both very tall. And there
was the drunken lady who told us about the nice ghosts
which haunted the stairs in her house.

Don always remembered one farm which he thought had
possibilities. It was out in the wilds of Wiltshire and up a
very narrow country lane. I went into the house while Don
examined the out-buildings. Don became acutely conscious
of a dreadful smell. He opened a creaking door and the
smell became even worse. The owner suddenly appeared
without warning, causing Don to jump six feet in the air.
He explained that this was a maggot farm . . . a very high-
class maggot farm.

The owner proudly shovelled up a heaving load and said:
"Now there's a fine class of maggot for you." The remain-
ing maggots were rustling and squirming in the background among dead sheep's heads.

We saw all sorts of properties including The Shady
Fragrant Tearoom, which conjured up tantalising ideas, but
our first serious 'find' was Elmley Castle near Pershore in
Worcestershire, not too far from Birmingham. Queen
Elizabeth I reputedly slept there. The house was huge and
shabby, the hall was like a giant's kitchen in a pantomime
and the enormous front door was propped up by a boat.

Elmley Castle had distinct possibilities and we decided to
apply for planning permission. The locals objected to our
plans and, after a while, so did we. We discovered dry and
wet rot and abandoned our application.

Don resigned from the zoo to give us more time to concentrate on our quest. One advertisement in The Field
caught our eye. We rang the estate agent and told him
about our bird zoo idea. He was so intrigued that he asked
us to consider an alternative and sent us the details of a
property which wasn't yet on the market. It was called Rode

Manor and was near Frome in Somerset.

Don and I first arrived at Rode Manor on Good Friday, 1961. It was being run as a market garden by an elderly couple. On the drive down Don made it clear that he had grave doubts about Rode. He felt a place of 17 acres would be too big for us to handle. I tried to be positive.

The Manor itself had been pulled down and, apart from the area near the lake where the owners did their market gardening in one of the walled gardens, most of the 17 acres - once the pleasure grounds of the Manor - was a jungle. But it proved to be a jungle with far more potential than anything we had seen so far.

As we drove up to the car park, a solitary cuckoo called. "That clinched it for me" Don told me later. "It was a message telling us that here was the place." And whilst proof reading this book I heard the cuckoo, the first time for several years.

We drove back to Birmingham discussing what we had seen. We kept talking about the beautiful back-cloth of wonderful trees, the grounds which offered us a chance to restore them to what must have been their former glory and, too, the potential of this part of England.

Bath and Longleat, two of Britain's major tourist attractions, were just a few miles away. Bristol, capital of the West Country, was less than an hour's drive. And, too, Rode Manor was perfectly placed, just far enough away from the main Bath-Southampton road to be peaceful and just near enough to make it easily accessible to visitors. Rode was the place for us, we agreed.

A very good friend of ours, a solicitor, helped with the initial legal matters. We applied for planning permission for our zoo and it was granted, subject to creating more car parking space in what had once been an orchard.

We got to know the owners who invited us to stay in part of the old house so that we could find out for ourselves if it would be suitable to make into a home of our own.

He had been a bank manager and had recently suffered a heart attack, which was why they wanted to sell. She was a difficult woman who drove everyone in the neighbourhood nearly mad. The two of them lived a very basic life. They had converted part of the old Manor staff dining hall into a bungalow, taking up the original wooden floors and replacing them with concrete. The walls sparkled with crystals from condensation and were painted a dirty blue except for the bathroom and loo, which were a garish bright red. They promised us the house was never cold.

The kitchen boasted a range called a Yorkseal, a sort of mini-Aga which was only lit once a week. The only other equipment was an electric kettle and an electric ring. The food was spartan - for breakfast, the occasional bread and marmalade over a cup of tea, lunches of cold meat or cheese and for dinner a sort of corned beef hash. The menu never varied except for the rare occasions when the Yorkseal was lit.

Our bedroom was equally austere and, at night - to Don's delight and my horror - bats flew in and around the room.

We learned about the history of our new home. The manor, known as Northfield House, had been built by a family called Andrews in the 18th century on the site of an earlier property. Several generations of the Andrews family lived in Rode between 1737 and 1796 but by 1822, the records show, the owners were Henry and Thomas B. Pooll. They owned about 10 acres of surrounding land and extended the house. The 1822 plan shows "pasture, fish-ponds and plantation" surrounding the manor and the annual rent, payable to the local rector, was £2 15s and 1 penny (£2.75 in today's money).

In the early 1870s the property was left to a relative, a Mr Langford, on the condition that he changed his name to Batten-Pooll. At the same time the name of the house was changed to Road - later Rode - Manor. The Batten-Poolls were great gardeners and it is said that a large number of trees were given as wedding presents and planted by them in their mainly formal gardens. Some of the wonderful trees visitors can see today date from this period.

Captain Walter Batten-Pooll, a son, died in 1954 and the property was put up for sale the following year. Rode Manor's decline was short, sharp and brutal. Lead vanished from roofs, trees were taken out and most of the house was levelled. All that survived were the stable block, old kitchen quarters and the bothies. The elderly couple's market garden began in 1956.

We moved to Rode properly on September 12, 1962 with two cars, some pigeons, Friday the cat and Nurse. Nurse had been Nanny to my two sons and we decided that she should be crowned as Queen of our gift shop . . . once we had one.

We spent our first night at a nearby hotel, Woolverton Grange, which the owner only opened when he felt like it. Luckily he was in a hospitable mood that evening so we had somewhere to stay.

Two furniture vans arrived from Birmingham the following day and I couldn't help noticing that the two men in charge seemed to spend an awful lot of time sitting down while Don read the paper. I felt I was the only one doing any work.

When the vans were finally unloaded and our belongings piled into the garage and the bungalow lounge, one of the drivers examined our new home, shook his head and told me: "Six months of this place will put 10 years on your lives."

There wasn't room for us all in one building so we used part of the old house which had once been the kitchen and bungalow to accommodate Don and I. We inherited a gardener who lived opposite the bungalow in the old stable block. Between these three buildings we had just 7 kilowatts of electricity, little enough for both heat and television too.

Not that there was much time to watch TV.

Don set himself a deadline. He was determined to open to the public by the following spring. I was just as concerned about making a home. Don suddenly decided he must have curtains in the furniture-choked lounge. I ended up hanging sheets over two sets of French windows. The walls badly needed attention and were in such a poor state that I had to have them lined with Kotina and then papered.

We rolled up our sleeves and got on with the job. Whatever happened, we were going to be open next spring.

When I wasn't helping paint and decorate the house and the bungalow, my main task was to dig and clear the stream that lay choked by undergrowth at the lower end of our grounds. Don also got Andrew and I to clear the rushes on the little lake - we complained that it was nothing less than slave labour! We made some impression on the overgrown lake as workmen dug away at the muddy bed until several began to complain of bad backs, including Don himself.

It was by now clear that we weren't getting on quickly enough, so we decided to hire a dragline, a machine at the end of a crane which was dumped into the lake and scooped up several tons of mud at a time. The mud was loaded into tipper lorries and emptied over the foundations of the Manor.

The dragline proved a godsend. Before we had completed the operation, we had shifted something like 4000 tons of the mud which had accumulated in the lake over the years.

I wasn't the world's best dragline operator but I thoroughly enjoyed my time at the controls even if others didn't. It had so many knobs, levers, brakes and so on that it wasn't the easiest machine to handle. Once I had successfully picked up a bucket of mud I always seemed to manage to swing it the wrong way so that the bucket ended up in a tree. The lorry drivers moving the mud were in hysterics about the activities of the stupid woman trying to cope with the complexities of a dragline.

Right, I thought, I'll show them all - I'll do it properly this time.

I concentrated as hard as I could, carefully followed the instructions, swung the bucket and let the load go. The men were still laughing, but not for long. One turned towards me and he was covered in gooey, sticky mud. So I had the last laugh.

The mud was dumped in large piles onto what is now our tea lawn. It was left in large piles which were later levelled by machine. It proved wonderfully fertile when we seeded and turfed it.

The stream began to flow more freely and the lake and pools filled up nicely, making fair-sized stretches of water which began to look very attractive. Because the water fowl we intended to introduce are vulnerable to foxes, stray dogs, cats and other predatory animals, we erected fox proof fences all the way round the lake enclosure.

While all this back-breaking labour was going on, we had to think about leaflets, guide books, postcards and all the rest of our publicity and information material. At least there were some amusing moments. One firm kept chasing us for the postcard contract and Don slyly asked the rep if he had any dirty postcards. The rep was not amused.

Now, at last, we could start introducing birds to the water. The first were a pair of Black-necked Swans followed by common Shelducks and then Don's favourites, Mandarins and Carolinas. Soon the water was full of life and interest as it rippled with the movements of the birds and sparkled with reflections of the lovely colours of their plumage.

We built aviaries and enclosures in other parts of the grounds and orders were placed with specialist importers and animal dealers for all kinds of parrots including macaws and cockatoos.

We were both very keen on ornamental pheasants. We bought in Golden, Lady Amherst's, Reeves and Silver

Pheasants and all the well-known species which are commonly bred in this country. Pheasants are very hardy creatures and can withstand even the most extreme English winters, coming as they do from the mountainous areas of the Far East.

My journeys across the country to pick up birds for our growing collection were a welcome respite from all these labours. One day I drove up to Birmingham to bring back a blue and yellow macaw named Raucous who went on to become the first father of many of our macaws. He must have been 3 or 4 years old and is still alive and with us today.

Lofty and Rosy, two more macaws, were also in this group. You will read more about them later.

Friday the cat, who had come down from Birmingham with us, had to be watched more and more closely with the birds. He had a nasty moment when he caught a mouse. He was trying to put it down to eat but was prevented by a gang of peacocks who surrounded him.

He went missing one day. We searched everywhere for him. And then I had an inspiration. I telephoned the greengrocer from Bath who delivered unwanted, surplus fruit which we could feed to our birds.

"Oh yes, Mrs Risdon, we do have a strange black cat here." Friday had climbed into the fruit lorry to have a look around and been trapped on board when the vehicle left. It was a great relief to have him back and he lived until he was 13. Since then we have never been able to have a cat because of our birds so we have kept dogs.

Things were taking shape and the grounds seemed to come to life after years of neglect. This had once been a place of great beauty with its formal gardens and rich ornamental plantings. Now the birds were bringing colour, movement and sounds all of their own with the squawks of parrots, shrieks of the pea fowl and noise of the toucans.

Don's only regret was that, inevitably, the natural wildlife would begin to suffer. Rode Manor had been an unwitting sanctuary during its neglected days and had become home to many creatures like grass snakes, slow worms, glow worms and beetles. Our birds quickly snapped up these tasty morsels.

It was important to set up a successful gift shop to help increase our income. This meant very careful stocking to ensure that we were selling items the public would buy. In the early 1960s the only worthwhile large Gift Fair was at Blackpool. Today's many fairs were still a long way off. Thankfully, the word went around and reps began calling at Rode bearing suitcases full of samples. All of them had to be seen just in case they offered the sort of goods we needed.

We also knew that good, professional catering would be essential. We had no cafeteria so we decided to serve light refreshments out of a window from the adjoining hall.

Those final months, weeks and days before we opened were mayhem but, thankfully, friends offered to come and help us with all the various tasks so that we could present the public with a complete picture of all we so longed to achieve.

The winter of 1961/62 was particularly harsh and it needed all our efforts to cope with the snow and bitter conditions. Yet work never faltered and I sometimes wonder how we did it all.

But spring came at last. We called in a photographer to take some shots to be used as postcards. He came on a particularly dreadful day and his pictures proved to be completely unusable. The penguins seemed to have pink spots and the rest of the birds looked terrible.

I decided I would take my own photographs from then on.

Our opening day was drawing ever closer . . .

CHAPTER FIVE

Early Days at Rode

W e held two open days before our official opening ceremony: the first for local people and the second for experts. The villagers had, of course, been very curious about what was going on at the Manor and we felt it was a matter of courtesy to invite them all to take a free look at what we had achieved since we arrived. They came in large numbers and seemed to be duly impressed.

We then held an open afternoon and tea party for Zoo Directors from across the country who had gathered for their annual meeting at Bristol Zoo.

Our distinguished guests were full of praise and some of them measured up our fences, made detailed notes about our aviaries, took photographs and generally picked our brains. It was only afterwards that we heard some had shaken their heads sadly and said: "They'll never make a go of it - they are far too far out in the country."

We were much more optimistic.

We needed a celebrity to perform the opening ceremony so we asked the famous radio and TV personality Maxwell Knight, who we had met at Dudley, to do the honours.

Max arrived on May 23 1962. We had a large gathering of Press, family and friends. The sun shone and the opening went splendidly. Just after Max stepped down from the platform, my mother whispered to me: "But he hasn't declared the bird garden open!" He handled the omission perfectly, returning to the platform among gales of laughter.

"I now declare these gardens open" a smiling Max said loudly.

Maxwell Knight performs the opening ceremony on May 23rd 1962.

Max came back to see us many times over the years and it was only after his death that I discovered that this good-natured naturalist was, in fact, the real "M" from MI5 before he found fame as a broadcaster.

Now, at last, we were in business and our daily routine changed dramatically. I shall never forget the sight of the first coachload of visitors to Rode. As the coach swung into the car park and the visitors descended, it seemed as if the vehicle had hundreds of people on board. We quickly learnt how to welcome the public and ensure that they enjoyed their day with us.

The building, painting and clearing continued and a couple of new jobs cropped up for me. The first was to make bird labels for the different aviaries. They needed to be substantial to withstand all weathers. I had not painted since my school days. I began by using stencils for the lettering and then attempted to illustrate the different birds in each aviary. At first we used wood for the signs but our free-flying macaws clearly felt that our wooden signs had been deliberately left out for them to chew. It was back to the drawing board - literally.

Our labels have advanced a great deal since then. We went from wood to metal and then formica and are now considering a new system. I developed my modest skills as a bird illustrator until Don took over the job and left me to design the maps showing the birds' natural habitats and to varnish each sign.

My second new task was to organise the catering. We desperately needed a commercial coffee percolator but simply couldn't afford one. Happily, the coffee rep who visited us offered to lend us one provided we bought his brand of coffee. I decided to produce home-made jam to pay for the machine and, after two years and 300lbs of jam, the coffee machine was ours. It is with us to this day, but in storage, having been replaced after giving us such sterling service.

It quickly became clear that we really couldn't go on serv-

Hard at work painting the umbrellas.

The completed cafeteria with guests enjoying cups of tea.

ing light refreshments and ice cream through the windows to visitors outdoors. We had set aside an area which had once been an old but fairly large greenhouse but it rained so hard one day that creating a covered area became a priority. We decided to convert a tiled section of what had been part of the kitchen into a cafeteria with tables and chairs indoors and outside.

I went to various sales and bought metal tables and wooden chairs. The tables cost £1 each and the wooden chairs just 2s 6d (25p). Except for the white tables, we set to and painted everything - including umbrellas - in bright colours. The effect was very colourful.

My duties also involved marketing the gardens. In those early days, we needed to do as much advertising as possible. The tourist industry as we know it today barely existed. Regional Tourist Boards, Tourist Information Centres and tourism publicity workshops were all things of the future.

We did our own mail shots to groups we thought might be attracted to Rode including schools, Women's Institutes, Townswomen's Guilds and hotels in the region. I well remember struggling with some 4000 postal items and stamping them by hand. It took forever. Today we simply put them in large bundles and take them to the Post Office for franking.

Don concentrated his efforts on the birds. As you read earlier, Don had always loved pigeons and he introduced pouters, croppers and blowers which love to strut and bow to each other. Don adored watching them flying around, gliding and parading. With their different coloured markings, they make a change from fantails or white doves.

Then one day he noticed that some of the pigeons had oily patches on their backs. He couldn't understand why until he spotted some birds vanishing under cars in the car park. They were after the road salt that had become encrusted under the chassis during the winter. Don bought a cattle salt lick and the problem was solved.

Our next winter was the bitterly cold one of 1962/63 when Britain came to a halt for days on end because of huge snowfalls. Our drive and the roads around us were impassable. We cobbled together a sort of snow plough by lashing old wooden doors to the front of the Land Rover and Andrew and I shovelled great piles of snow to clear the drive. At times our part-time keeper had to clamber over railings and snowdrifts to get to work.

The birds needed constant attention during this crisis. I recall trying to catch a white peacock in the snow - it was almost impossible to see. The temperature fell to minus 14°C and our kitchen was always full of cages of birds needing warmth and food. Some were suffering from frost-bitten feet and had to be hand-fed. We built straw houses for the flamingoes but lost a couple who died of aspergilosis from the straw. It was a white nightmare which lasted for weeks, but somehow we survived and so did most of our birds.

We have often had to battle with the weather over the years - including snow and ice in April. There was a dreadful moment in 1983 when we suffered a sudden and very hard frost and lost all our electricity. We had 14 deep freezers, most containing food for the birds, and had let several outbuildings for an aquarium. Mercifully, the Army came to our rescue. Units from Warminster arrived with generators during a critical period when we were without power for more than 2 days. Today we have increased our winter quarters for the birds and have heating in some of the aviaries.

It must surprise many people that birds from other, warmer, parts of the world can quickly become acclimatised to the vagaries of the British weather. In fact, they are almost always better off out in the fresh air than in a hot, stuffy atmosphere indoors.

Extreme winters are probably hardest on the keepers who need to keep water pots filled and unfrozen and, above all, prevent locks being jammed by ice. The answer to the latter, we soon discovered, is to insert glycerine into any working parts. It does the job perfectly.

If thick snow falls on the aviaries, it has to be moved as quickly as possible. I have found that the best way to shift it is to don good waterproofs, go into the aviary with a broom and push upwards. Almost inevitably I end up looking like one big snowman!

Droughts and floods have been our other weather problems. When it comes to coping with droughts, we are lucky enough to have our estate water supply as well as mains water. Floods have caused some havoc over the years but the drainage around the grounds inherited from the Rode Manor days has helped a lot.

The greatest damage has come from gales. I will never forget that terrifying day on January 25th 1990 when the great hurricane hit Rode at lunchtime.

Trees were falling all over the place, bits of plastic from the cafeteria and gift shop roofs were flying everywhere followed by tiles and a TV aerial attached to a chimney pot. Throughout this horrifying disaster the keepers and staff did a wonderful job bringing the birds to safety.

We lost more than 100 of our beautiful trees and to this day I shall never know how no-one was killed or even injured. Our Guardian Angel must have been working overtime that day. We only lost two birds and had one injured. Staff unable to get home because of roads blocked by fallen trees worked very hard to patch up the broken foxproof fence around the lake to protect the birds in the waterfowl enclosure.

We had to close the Bird Gardens for weeks to clear up the mess. We put up notices all the way down the drive informing the public, but still some people came. One car arrived and the driver demanded to be admitted. We told the group that we had moved a lot of the birds for safety's sake but he was adamant - he had only come to inspect the damage, he said.

Another car-load, with children, were discovered wandering around among the Danger signs warning people that the trees were unsafe and could fall at any minute. There was even a caravanning group who bought out tables and chairs and began having a picnic!

Over the years we have learnt that TV and radio warning services for emergencies are a great help. There is just a tiny minority who take absolutely no notice at all. They are deaf to warnings and ignore danger signs. Their attitude seems to be: "Oh, that can't possibly mean me."

We began our second year with great hopes. Our name was becoming better known and on really busy days there was the odd remark that we must be making a fortune - very far from it. Some visitors even said they might consider starting a bird garden of their own. I rather think they imagined that all we did was to stand on the lawn and scatter bird seed. If only they knew! They never saw the 101 chores that go with running a business like ours, from organising car parking to cleaning the loos.

Don once complimented me for being adaptable and flexible. I used to laugh and say that he'd found his perfect slave. But it is true that when I'm confronted with a problem, I simply ask myself: "What am I going to do about this?" I usually work out the answer.

Lots of people are attracted by the life of helping run a bird garden and we have had many applications for jobs over the years. I really rather think I missed a treasure by not taking on a 32-year-old clerk who wrote: "I am broad-minded, progressive, flexible, creative, imaginative, resourceful, realistic and have had 20 years on a variety of activities involving organisation, leadership and personal achievement." All of them qualities that you need running an operation like ours.

On several occasions we had people who swore it was their life's ambition to work in such an ideal place . . . and then got cold feet at the last moment when we offered them a job.

Some of our hardest workers were our many friends who came with their families to lend a hand both before and after we opened. We put them up and in return they gave us wonderful service, digging, hacking and clearing stone from the old manor house. They built walls, made paths, built great bonfires of scrub and brambles and proved a godsend. We could never thank them enough for all they achieved.

One couple from the village were full of enthusiasm until we found the husband staggering back from the woods where he had been clearing undergrowth. He demanded a large whisky and swore he wouldn't be back until it was too dark to do any more work.

Marlis, a young Swiss girl who was the daughter of a friend, proved a treasure when she spent a working summer with us. She was extremely keen and took great care of the birds. She had a particularly soft spot for Sigee, the Bare-fronted Sulphur Crested Cockatoo that we had been given. We even found her knitting him a waistcoat to keep him warm. On another occasion, she was found bringing lambs in and covering them with blankets because "the sheeps were getting wet."

Sigee was just as fond of her and liked following her around. One day Marlis was helping to paint the cubicles in the Ladies toilets and Sigee was walking along the tops of the cubicles and inspecting her handiwork. Marlis didn't hear the visitor who entered and locked the door . . . but she certainly heard the horrified scream and bang of the door when the lady ran out after Sigee suddenly appeared above her.

I'll never forget another incident, again in the Ladies . It was a busy Sunday during our third season and I was about to walk through our small gift shop when the lady in charge called me over.

Sigee, the Bare-fronted Sulphur Crested Cockatoo at play among lawn clippings.

Betty and friends in the Bird Gardens

"I think you had better come and help, Mrs Risdon" she said, telling me that someone at the pay box had reported that it sounded as if someone had broken a leg in the Ladies. I hurried over, wondering what I should do. Call for a doctor? Order an ambulance?

A visitor explained that she had found broken seats in the Ladies. I ran round quickly. Don and our manager Mike Curzon stood by while I went in to see what had happened. Inside there were two broken seats lying on the floor to one side. I got them out of the way and then took a closer look at the scene.

Slowly it all began to make sense. A rather heavy lady had clearly become impatient trying to pull the chain which would not work properly, so she climbed on the seat to put the chain right. In her struggle she must have pushed the top of the cistern over the top of the adjoining cubicle and then slipped, breaking the first seat. The cistern had smashed the loo in the next cubicle . . . thank heavens no-one was in there at the time!

When I came out to tell an anxious Don and Mike, I simply couldn't keep a straight face. I laughed so much that tears poured from my eyes. It took ages before I could explain - and then Don and Mike started laughing so much that it was several minutes before we all managed to pull ourselves together and start getting the Ladies back to normal.

A distinguished visitor to Rode in the early years was the late multi-millionaire Paul Getty, then said to be the richest man in the world. He came to make the South West regional draw in aid of the World Wildlife Fund.

Mr Getty swept into Rode in a swish silver-green Cadillac nearly three-quarters of an hour late. There were a lot of Press reporters and photographers who had been firmly told that no-one should ask Mr Getty about money . . . or anything else, for that matter!

He made a very short speech to the waiting journalists,

saying quietly that he had come because he was very interested in wildlife.

"I'm here because they asked me. I have a small zoo at home in California which has some deer, bison and lions. No-one can accuse me of breaking and entering. Anyway, it gives me the opportunity to visit your beautiful West Country again". And that was that.

We showed him around the grounds and he told me he found them beautiful, which was a nice compliment. He sat with Don and I over a glass of champagne and a buffet lunch. I found him rather aloof but Don and I were delighted with the publicity his visit attracted and there were pictures in all the local papers.

Meanwhile, as our collection of birds increased, so did our breeding programme. Conservation was becoming more and more important to protect tropical birds. We developed invaluable contacts with other bird gardens and zoos, lending or borrowing birds on breeding loan.

We achieved a series of "firsts". For instance, we were the first centre in the country to successfully breed Umbrella Cockatoos and Scarlet Ibis. We are very proud of these achievements and my thanks go to Mike Curzon and the keepers who have done so much to help us with this important work over the years.

Since the advent of the Wildlife and Countryside Act in the 1970s, we have conformed to the Zoo Licensing and Inspections rules. In fact Don played an important part in helping frame the Act together with other Zoo directors. He would disappear to London for meetings, often in the House of Lords, and would arrive home exhausted and glad to be well away from the rush of life in the capital. But it was worth the effort once the legislation had been passed.

One rule which we imposed ourselves right from the start was a ban on dogs. The main reason - apart from the obvious risk of dog attacks with so many birds at liberty for the public to enjoy - is the danger of disease. I painted a very large sign for the car park which showed a very ferocious peacock chasing a dog and warned: "For its own safety, please keep your dog in the car."

The ban does not go down well with some visitors and at times they try to smuggle dogs into the grounds. They either conceal them in a bag or try to slip them past the pay box attendant at the busiest times.

When owners do arrive with dogs in their cars, I always insist that they park their vehicles in the shade and leave a window or windows slightly open so that their pet has plenty of air I do realise that some people feel unhappy about this with so much car vandalism these days. I just wish dog owners would leave their pets at home if they want to enjoy our bird gardens - it would be far simpler for everyone.

I hate having to put up lots of signs - like "Please don't chase the birds" and "Please don't pick the wild flowers - let everyone enjoy them" - but they really are necessary. You would hardly believe what some people will do. I have seen people carrying bunches of primroses, cowslips and even ordinary garden flowers. We always confiscate them.

Not long after I had been planting in a small bed near the toilets, I was walking up to the top car park when I saw a visitor with an armful of my plants. I asked where they came from. He claimed he had bought them at our plants-for-sale stall, which was ridiculous since we only sell clematis. He angrily handed them over with a volley of abuse.

One spring-time the tulips were just starting to appear in the main beds by the tea lawn and the daffodils were just beginning to flower when I saw three children jumping about in them. I asked them to stop and go back to their parents. They did stop - but only for a couple of minutes.

When they began jumping on the beds again I asked their parents to tell them not to damage the plants. Their response? They told me to mind my own business, that they had paid their entrance fee, that it was up to them what they did and that they were never coming back again. I'd love to know how they would have felt if we had damaged their garden. It is all very sad.

We have made it clear from the start that unaccompanied children who misbehave might be sent away. One lady came to me and protested about this. "You can't do that" she said. "If you send a little girl out of the gardens she might be raped."

Honestly, when you are feeling fed up with these sort of complaints there is no greater tonic than someone saying - as they often do - "I do want to thank you for giving us such a wonderful day. We have enjoyed ourselves so much and want to congratulate you on all you have achieved." Suddenly, life takes on a new meaning. All our efforts seem worthwhile after all.

Life here has many funnier moments. One day a man came to the gift shop and asked me: "Have you any manikins?"

"Yes, sir, you'll find them in the aviary near the monkey".

He looked at me and said curtly: "I said manikins".

I could feel the tension mounting as I told him once more where they could be found. By now he was furious and retorted that I should have known he was asking for a brand of miniature cigars!

On one particularly busy day I was manning the pay box when I was asked by a foreign visitor if we had any Hamburgs. I called out to the gift shop to ask how many of the Silver Hamburg Bantom eggs we sold for breeding were left. I sent him round to the proper window but he sailed past into the cafeteria where he was offered ham-flavoured crisps. He came storming back angrily to tell me that all he wanted was a hamburger!

One visitor who didn't know who I was told me that our bird garden was owned by Peter Scott.

"No, honestly, it's privately owned by my husband and

myself."

"Nonsense. I know for certain that it is owned by Peter Scott." That was the end of THAT conversation.

One forlorn party arrived turned to me with disappointed faces. "But where are the lions? Oh no, isn't this Longleat?" I do hope they enjoyed their visit . . . even if we couldn't offer them lions.

Don and I were sometimes asked the silliest questions, like "What do you do with the birds in winter? Do you let them die off and buy a new lot in the spring?"

We love meeting the public. The vast majority are a delight to welcome, but there are times when it really is rather hard not to tease. One visitor who was very impressed with all our free-flying tropical birds asked me why they never flew away.

I asked: "Have you seen me waving my arms to and fro?"

"Yes, but why?"

"Well, you see, it rather depends on the weather but if the sky is grey, I use grey cotton, but if it is bright blue, I use blue cotton . . . tied to the birds' feet. That way I can control them."

We also receive some very unusual letters and phone calls.

"Dear Sir,

Please can you help me. My budgie, aged 4 years, fell into my husband's dinner yesterday. I sponged his chest and cleaned his feet with luke-warm water, slightly soapy, and put him back in his cage.

After a while I let him out again but he just flopped off his landing stage and fell onto the radio. He has never been a strong bird on the wing but this is definitely his worst performance. I rang the vet who told me to keep him in his cage for a week to let him rest.

I am terrified that he may drop down again or hit something or kill himself. Apart from this he is quite

*normal, eats well, is bright and intelligent and I love
him dearly.*

*What can I do? Can I bring him over to Rode for
you to see. I am very willing to pay for your help.
'Ecclesiastes', so called because he refuses to talk, has
always loved his freedom and all his toys dotted about
the room. He cannot stay shut in a cage.*

Please HELP."

We were able to reassure Ecclesiastes' worried owner that
their poor pet had probably been suffering from nothing
worse than slippery feet from the soapy water.

Don was often at the receiving end of phone calls about
birds.

"I've got a bird in my garden, is it yours? Is that Mr
Risdon?"

"Yes".

"What sort is it?"

"Oh! I don't know but it's a foreign one."

Don (calmly): "How big is it?"

"Quite big."

"Bigger than a robin?"

"Oh, much bigger than that."

"The size of an albatross?"

"Oh no, it's just an ordinary size."

"Is it a pigeon?"

"No, it isn't one of those."

"What colour is it?"

"Sort of fawn and fluffy - you know."

"Has it big eyes?"

"No, I can't see the eyes."

"Has it long legs or big feet."

"I can't see its legs or feet."

"Has it a hooked beak?"

"Oh! No, it's a flattish sort of beak . . . now you're getting me all flustered."

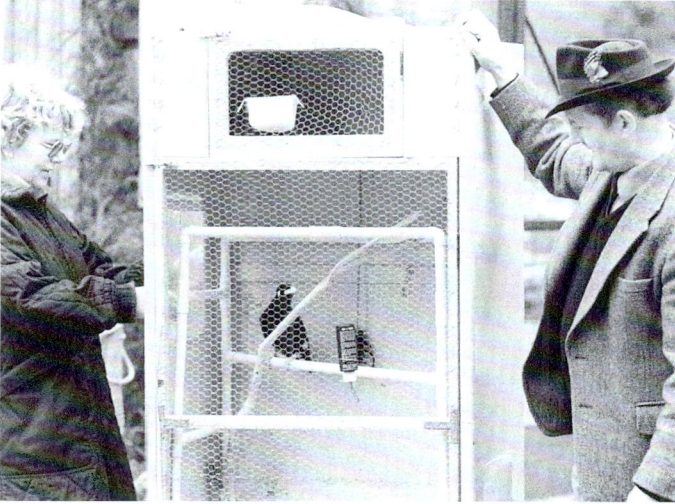

Bertie in his wardrobe.

"Well, Madam, I should leave it where it is. I don't think it comes from here."

Don assumed that it was that rarest of birds, the eyeless, legless, fluffy fawn tree creeper.

Or another caller:

"Have you lost a bird from the bird gardens?"

"Yes, have you found one? If so, what sort is it?"

"Yes, I have found one but if I tell you what sort it is you'll say it belongs to you."

"I think you had better get in touch with the police and tell them what you have found. Thank you."

And another:

"Please can you help me? My parrot has just bitten me. What shall I do?"

"How bad is the bite and have you had a tetanus injection? If not, may I suggest that you go to your nearest hospital and tell them what has happened."

The same caller rang back a few weeks later.

"Please can you help me. You must think I'm mad - but my parrot has just fallen into the emulsion paint. He is the one I telephoned you about when he bit me."

"You went to hospital and did they give you an anti-tetanus injection?"

"No, they said I would not get rabies."

Don offered some helpful suggestions about how to deal with the poor paint-sodden parrot.

But just occasionally came a letter or call which really did turn up trumps. Johnny Morris passed one on to me.

It read: "I have a problem. I have 3 cats, 3 lizards, tropical fish, 3 dogs and Bertie, a Javan Hill Mynah. He does not talk but is driving us mad with his imitations. He lives in a converted wardrobe. The top is his bedroom and in the rest he insults our dogs and cats and, if they come too close to him, he has a go at them".

We offered Bertie a home. He arrived plus single wardrobe with bedroom on top. Bertie lived with us for some time. We put him out with our other mynahs once he had been weaned from his wardrobe. As for the wardrobe, we used it for storage.

CHAPTER SIX

Natural Enemies

We allow our birds at Rode as much freedom as we possibly can . . . but we have to guard against natural predators from the surrounding countryside. One of our first precautions was to create a fox-proof fence around the lake where the water birds live. At the time we simply couldn't afford to pay for a protective fence around all of our 17 acres.

When a lot of the pheasants and pea fowl began nesting, we had to take special measures. These birds are safe when they are roosting up trees and out of reach of foxes and badgers and other predators, but once they lay eggs they are at great risk. They nest on the ground and are terribly vulnerable to any bumbling badger that goes by or any fox on the look-out for an easy meal. We lost many birds in the early days.

We overcame this to some extent by using the old game keepers' trick of hanging lights in the bushes by where the birds are sitting. This seemed to scare most - but not all - of them away and some damage continued.

It was becoming apparent that although we could barely afford to put up a full boundary fence, we had to do something. Lack of protection was costing us far too much in the loss of birds.

So we improvised and came up with a far cheaper solution to what was becoming a real headache. The greatest expense in fox-proofing comes if you start digging trenches for your fences. It's a costly, labour-intensive job.

Don thought about this problem and came up with a far cheaper - and equally effective - solution. He experimented and found that if he brought the wire fence down to ground level and then bent a three foot apron of wire outwards from the bottom, it would do the trick. Once in place, spread outwards and stamped into the ground, the apron quickly became covered with grass, forming an impervious mat of turf and wire.

When a fox tries to dig under the fence, its paws strike the hidden wire and it gives up. Foxes run along fences with their noses down against the point where the wire reaches the ground, searching for vulnerable gaps. Luckily, they don't have the sense to retreat three feet back from the fence and start digging there.

Putting up this fence was one of the best investments we ever made and the number of losses fell sharply.

Now we no longer had to shut our birds in at night or go around the grounds lighting candles inside tins to frighten off foxes, a tiresome, time-consuming chore. We could safely leave the birds to find their own nesting and roosting places.

Unfortunately rabbits do have the sense to start burrowing well back from the fence and they often manage to breach our security. A fox may follow them and try to use the rabbit hole so we have keep a close watch on the boundaries to ensure that any holes are quickly filled.

A foxproof fence also has to have an overhang to keep stray cats and dogs at bay. They run along the upright and cannot mount the fence. These fences are not proof against mink or weasels - for that you need a much finer mesh.

Badgers are often portrayed as bumbling sorts of creatures which wander around and present no threat, but if a badger trundles across a sitting peahen, turkey or pheasant he'll have it quick as a flash, kill it, tear it to pieces and then eat all the eggs or chicks. This has happened several times and we are always on the look-out for badgers. These powerful animals are great diggers and are much more likely to dig under a fence than foxes.

They can, all too rarely, provide moments of amusement. Just a few weeks ago, the chap cleaning the staff loo came across what he thought was an abandoned coat. On closer inspection it proved to be a badger curled up in a corner. He had to use a broom handle to remove the unwanted "coat".

Among our other pests are grey squirrels which get through the wire of the aviaries and help themselves to the inmates' food. They are also a menace in the gardens, particularly with bulbs. Their favourites are tulip bulbs which they dig up.

Mink can be devastating. We have a little stream and, very occasionally, a mink will get into the waterfowl enclosure. The first sign is decapitated corpses lying around.

Mercifully, this has not happened for many years but I well remember the day we first discovered that we had a mink in the enclosure. One of our keepers spotted it sleeping in one of the duck nest-boxes. We went down with a shotgun and tried to shoot it, but without success.

We had an alsatian guard dog at the time. She was extremely brave and when the mink shot out of the nest-box into the water, she spotted it when it broke surface for air. She grabbed it, dragged it out of the water and killed it after a heroic struggle. We were all very relieved that this menace had been dealt with.

Stoats can be a threat too although Don once saw an extraordinary sight in which a stoat was seen off by a rabbit. He was sitting watching his beloved pigeons when he suddenly noticed something streaking across the lawn in front of him. He had his field glasses with him at the time and watched a rabbit being pursued by a stoat. He had no doubt at all that it was a stoat because of its colour and movement.

Don was expecting a kill when, suddenly, the rabbit turned on its heels and began to attack and chase the stoat. If Don hadn't seen this with his own eyes, he would never have believed it, he told me.

Herons can be quite a nuisance. Don kept fish in the pond behind our bungalow and protected them by hanging metal cats' heads by the pool to scare them away. A heron could clear a pond like ours by coming early on summer mornings before anyone is up and about, eating the smaller fish and maiming ones which were too hard to swallow by spearing them with its sharp beak.

The herons were also a pest when we were feeding sprats to the penguins. They would hang around in the field next door hoping for a good meal when the keeper had left. We managed to overcome this by hand-feeding our penguins.

But these natural pests pale into insignificance compared to the worst enemies of all . . . bird thieves. I shall leave them till last . . . it is not a subject I enjoy writing about.

CHAPTER SEVEN

Don's Favourite Birds

The bird gardens were running efficiently. The installation of fox-proof fences had done a great deal to ensure the birds' safety and, with the first, hectic period of creating and opening our bird zoo, we could now concentrate our efforts on a series of improvements to the grounds and continue to increase our collection.

Although my knowledge of birds was growing all the time, I was a comparative novice compared to Don. He was completely in his element, surrounded by so many of the birds he loved and admired the most.

Don was always being asked which was his favourite bird. It was an impossible question to answer - he had so many. But he did admit that the Royal Starling was very near the top of his list. This truly beautiful bird is elegantly built with long, slim legs and blue plumage above and gold-spun below. In Africa it is called the Golden Breasted Starling, a very good name. However, the colours of deep blue with a touch of green on the head and the golden breast are truly royal colours. No wonder it was given its regal name.

The Royal Starling's overall colour scheme is covered with a sheen which makes the bird appear to be cast in burnished metal. That is why this group are given the name Glossy Starlings.

Don had kept several of these birds in his aviary at Hove before the war. He had never succeeded in breeding them but had never forgotten their beauty. When we were offered some birds, he leapt at the opportunity. Thanks to major improvements in our knowledge of diet, Don at last managed to breed his own chicks. It was one of the very first successes in breeding Royal Starlings in Britain.

We had made a collection of ornamental pheasants from the very start. Don was particularly keen to establish our own flock of Golden Pheasants. We had the ideal grounds with plenty of woodland and undergrowth. Golden Pheasants come from the mountains of China and in the wild they inhabit rhododendron forests. Our soil here at Rode is alkaline, making it impossible to grow rhododendrons, so instead we grow laurel bushes.

The pheasants established themselves quickly and even began to rear young of their own. They are experts at hiding their nests and it can be very hard to find a sitting hen. It is said that a sitting Golden Pheasant hen will sit for 24 days without food or water rather than leave the nest.

If we do find a nest without a hen, we remove the eggs and put them under a bantam and rear them that way. We also keep a pen with a cock and several hens. When eggs are laid, we remove them, again put them under a bantam and later release the birds in the grounds. In this way we keep the stock going.

Golden Pheasants are different in temperament from their near-relatives the Amherst's and other pheasants - they are not pugnacious except at spring time. They seem to live in a sociable flock and when they do display, usually from the autumn onwards, they have a social display during which the cocks gather 12 or 15 birds in a group and show off. They run in circles around an immature bird or old hen. Their plumage in full colours makes a glorious sight.

In springtime their nature turns and they become territorial, the dominant cocks driving away the others. By the summer we manage to replace those that have been driven off and so keep our stock going.

We have tried having other pheasants like Swinhoe, Silver, Reeves and Amherst's run loose in the grounds, but some are far too territorial and aggressive in the autumn and do not put on the wonderful shows of the Golden Pheasants.

While Reeves look good they can become too used to humans and there is a danger of them attacking members of the public who want to feed them. The same is true of Silver Pheasants who can attack not only other pheasants but also children We also discovered that Silver Pheasants had a habit of forming packs to hunt the Golden Pheasants.

Don's dream of all our colourful pheasants wandering around the grounds simply did not work in practice except in the case of the Golden Pheasants. They are easy to look after and breed, but you do need to give them the right habitat with woodland and dense undergrowth where they can hide

Many people imagine that it would be rather a nice idea to have ornamental pheasants wandering around their lawn, but these birds need the reassurance of good hiding places as well as open spaces.

Don was always very keen on cranes and we have a fine representative collection. Our first cranes were Demoiselle and Crowned which were the species most readily available. Today we have breeding pairs of Demoiselle, Crowned, Sarus, Stanley - the beautiful Blue Paradise Crane from Southern Africa - Maguari Storks and Wattled Cranes.

Cranes are very elegant birds and are spectacular both in their appearance and their display. They are hardy and easy to keep and fairly easy to breed provided you have compatible pairs. However, cranes can be very quarrelsome and have to be kept in separate pens.

When cranes lay, we take the eggs and either put them under bantams or in an incubator because they do not make very good parents. If it rains or you have a poor summer, the parents have a habit of leading their chicks through long wet grass as they hunt for insects. The chicks then get wet and chilled and can all too easily die from exposure. However, when they are reared under bantams

Crowned Cranes . . . Don was always very keen on cranes.

they are taught to eat by their bantam foster mother.

The six-foot Sarus Cranes always have two eggs and hatch two chicks. Each parent takes charge of an individual chick. The reason appears to be that the chicks have a natural tendency to fight each other. We discovered this with chicks raised together by a bantam. One chick would almost inevitably kill the other. We now separate the chicks until they are older and the instinct to fight has died down. Remember the rhyme: "Little birds in the nest should always agree/And never fall out with the family".

Don's favourite was the Crowned Crane from Africa. There are two species - the Black-necked from West Africa and the Grey-necked from East Africa. The latter is slightly larger and is much more colourful and showy. It has an entirely different voice which is a short, often-repeated call.

We have bred both of these species at Rode. If they are hand-reared they can become 'imprinted' - too familiar with humans. This can pose a threat, so we put them in a

Head of Crowned Crane
Opposite: White Pelican

Lofty the Green-winged Macaw.
Previous page: Betty Risdon, 1998. (Photo: Jerry Richards,
courtesy of The Western Daily Press).

The grounds in summer.

A birds' birthday party.
Opposite: Moluccan Cockatoos.

Baby macaws, barely four weeks old.

Eagle Owls.

Don with wild Rainbow Parakeets at the Currumbin bird sanctuary . . . bliss for the dedicated bird man.
Overleaf: The lake in winter.

pen to prevent them causing any injury to visitors or keepers: a peck from one of these large birds can be very
painful. Another reason for keeping them in pens is that if
they were at liberty there would always be the danger of
them flying off and getting shot.

Imprinting also affects penguins if they are hand-reared.
Penguins are not very good parents so when we breed
them, they nearly always have to be hand-reared. The
chicks are taken when they are a few days old and fed on
chopped fish. It is very difficult to wean them back to their
own parents.

Penguins are charming birds when they are tame - you
can bend down and stroke them - but visitors are some-

South African Penguins

times tempted to let them out of the waterfowl enclosure.
Freed penguins often head for the car park where they look
as if they are trying to direct arriving and departing traffic.
Once we even saw a visitor pick up a penguin, put it in a
pushchair with a child and take a photograph.

Aviary birds are those which cannot be let loose naturally and either have to have their wings clipped or be kept in
enclosures. Our aviaries are as large and spacious as possible and we plant and turf them to make them appear like
mini-gardens. Each one is carefully landscaped and stocked
with a variety of mainly perching birds alongside attractive
pheasants like Tragopans. Grey peacocks and pheasants are

very happy in aviaries too.

We found that Hornbills are hardy and far easier to keep than Toucans. The latter have a tendency to live happily for a limited period and then suddenly die. Several Hornbills have attempted to breed in the nest boxes we have provided for them.

One of Don's favourite aviary birds was the Touraco. This lovely bird from Africa is coloured in shades of dark green, purple and dark blue with patches of white here and there. Their movements are extremely graceful and they proceed around the aviary with a hop, skip and jump rather like a squirrel jumping from bough to bough. When they fly they show flight feathers of a lovely shade of carmine. Touracos have nice, deep voices, rather like a cross between a pigeon and a crow . . . a sort of curious croak. The call is loud and far-reaching.

The best breeders are the White-cheeked Touracos. We manage to breed them every year. They live almost entirely on fruit and build a twig nest on a platform where they lay two eggs and usually manage to raise both chicks.

Another bird we acquired was the famous Kookaburra from Australia. These birds with their loud laughing cry are particularly popular with visitors. We have bred several Kookaburras. One was hand-reared and could be made to laugh quite easily by imitating the sound. It really is a maniacal laugh although the birds themselves tend to be rather wooden and sedentary.

Don had an amazing ability to know whether a bird was sick or not, even if it looked perfectly healthy. This even happened towards the end of his life when his eyesight was deteriorating. Before a bird developed noticeable symptoms of an illness he would often bring it in and care for it. He had great sensitivity . . . a sort of bird ESP.

He was quick to spot when one particular penguin was showing unusual signs. Normally if a penguin becomes sick, it quickly dies. This was something different. Don had the bird X-rayed and it was found it had swallowed an eight inches long rubber snake. We were selling these sort of novelties in the gift shop at the time. Obviously some child had fed it to the poor penguin. An operation was carried out to remove the toy and the penguin survived, complete with operation scar.

CHAPTER EIGHT

Fun ... and Friends

Running our bird garden has been hard work over the years but it has been worth every moment. We have had a lot of fun and made many friends . . . both feathered and human.

Collecting birds from airports, railway stations or bird centres in other parts of the country can often be hilarious.

Shortly before we opened, I travelled to Heathrow to pick up eight Cape or Blackfoot Penguins from the coast of South Africa. As I was leaving the Import Embargo area, I was stopped by a bleary-eyed pilot who was scratching his head.

"Excuse me" he said, "I had a rather rough night last night and I think I'm seeing penguins". I was able to reassure him that he was, indeed, seeing penguins . . . each one housed in a nice box from which they could look out on the world and bemused airline pilots!

Another pick-up was to collect several flamingoes whose legs were encased in nylon tights. This is a simple way of preventing any damage to their long, stilt-like legs, but it does look rather odd when you are standing waiting for an airport bus. As for the sight of their swaying heads in the back of an estate car - that is quite unbelievable.

I once went to meet our friend Gerry Kirkham who was bringing us a selection of birds from East Africa. Those were the days when it was still possible to import birds without a period of quarantine. It was late at night when I set out for Heathrow and well into the early hours before the birds had been cleared by Customs. I set off home with a fully-laden car and was flagged down by a police patrol

A car-load of Cuban Flamingoes arriving at Rode, their legs encased in nylon tights.

car. The two officers wanted to know what I was carrying.

I told them: "I've got a lot of birds and a baby ostrich." It was just getting light. You should have seen their faces - they had certainly never stopped an ostrich before.

Don and I were on our way to the Norfolk Wildlife Park one day when we ran into a road census. We were asked where we had come from, where we were going and what was the purpose of our journey. We explained that we were going to collect some birds for our bird gardens. On the way back we ran into the same census, but this time manned by a different girl. Don told me to keep quiet.

The same questions were asked and when she asked the purpose of our journey, Don kept a completely straight face and announced: "We're going to take both our rheas home". I fell about laughing as she blankly looked first at Don and then at the two heaving sacks in the back of the car.

Exactly the same thing happened when my son Andrew

and I were driving to Trowbridge to meet a train. We told the census people that we were going to collect a pair of nuns from the railway station.

We were stopped by the same people on our way home. "Weren't the nuns there? Did they miss the train?" we were asked.

"Oh they're here all right" Andrew said. "They're in this box - they are pigeons."

At Rode many of our birds became great friends and were great characters. One was a pelican we were given some years ago by someone who did not know what to do with him. One of his wings was clipped but he eventually moulted out his flight feathers and took to the air. We named him Boeing.

Boeing was a free spirit and roamed far and wide but always returned at penguin-feeding time for a snack of sprats. One day we received a complaint from a neighbour who had a lake which was open to the public. It transpired that Boeing had been landing on his lake, first eating his fish and later some ducklings. We had to clip his wings again and he took to sitting on the flamingo nests, frightening the birds away.

Another friend was a very tame Queen of Bavaria Conure. A parson friend came to lunch and was intrigued and charmed by the little bird which perched on his knees. The Reverend said he was going to startle his congregation that Sunday by telling them that he had had the Queen of Bavaria sitting on his lap!

Freddie Maggs was a great personality. He was a very handsome magpie with a wicked sense of humour. We had not had him very long when, at about five o'clock one morning, there was a plonk on my chest followed by a discreet cough. It was Freddie. It was his way of letting me know that he was up and about, so why wasn't I?

Freddie loved being with us and his greatest ambition was to fly into the bungalow and perch on top of a door, hoping we wouldn't spot him. It was always that discreet cough which gave him away. Freddie also loved hiding things in strange places - bacon down the back of a chair and, one night, a piece of stewing steak in my face flannel.

He once mischievously stole a girl's wedding ring which she had taken off to play with on the lawn near our house. Freddie nabbed the ring and flew up to our roof, teetering on the edge of the guttering and teasing the poor girl below.

She had a fit, shouting: "My husband will kill me!"

It took us quite a time to catch Freddie and reprimand him for being so discourteous once the ring had been restored to its rightful - and very relieved - owner. It was a very sad day when Freddie vanished. We never knew what happened to him.

Lonnie, an Indian Hill Mynah, was unforgettable. We first met when I had an early morning call from Trowbridge railway station. The man in charge was being driven nearly mad by a brown paper parcel addressed to Rode which kept saying: "I can see you."

Lonnie had a cockney accent and a wide vocabulary. He was put in an aviary and caused all sorts of misunderstandings. A couple of elderly ladies had only been in the gardens a few minutes when, pink-faced with anger, they demanded their money back, claiming that a nasty man was hiding in the bushes insulting passers-by.

It was Lonnie, who had shouted: "Hullo my darling. I can see you. You silly old baggage" at this very respectable couple.

Among the birds I brought back from the Tyseley Pet Stores in Birmingham shortly after we opened were three macaws. One was Raucous, a blue and gold macaw, who was the first to breed here. He's a noisy character, hence his name, and he is still with us, although now rather old. Rosy, a small scarlet macaw, made friends with Lofty, a red and green macaw. They lived in an aviary not far from the cafeteria lawn and when rain threatened, they would solemnly walk over to warn us.

Lofty earned his name because he made an annual flight to one of the highest trees in the grounds. He would sit there until sighted. He really could not fly very well so we had to talk him down from the tree. This became an art in itself.

"No, Lofty, the branch below you. Come on. You can do it!"

When, at last, we had managed to bring him back to earth, he was delighted and relieved not only to be on terra firma once more but also to be able to enjoy something to eat.

In later years, when we began to "sex" our birds, we discovered that Lofty was, in fact, a she and we believe she must have already been fairly old when she arrived and well into her fifties when she died. It is always sad to lose such a friend, and we missed - and still miss - Lofty.

Sigee, the sulphur-crested cockatoo who caused such havoc in the ladies loo when 'helping' our young Swiss friend Marlis, came to us in 1963. He had lived in a pub for 12 years and swore like a trooper. He was allowed to fly free sometimes but hated many of the staff here and would try to attack them. One of the ladies in the cafeteria wore a wig and if Sigee was loose he flew into the loo after her and the poor lady's wig was very crooked when she returned to the kitchen.

I am sorry to say that Sigee was stolen at the end of 1994 . . . but more of that later when I write about the wretched subject of bird thefts.

Usher was the first penguin we bred. He was reared in a beer barrel from the famous Trowbridge brewery Ushers, hence his name. He was raised alongside our manager's new-born baby son. We never really discovered if they were both fed minced fish! Usher became very tame and would allow visitors to stroke him. But if he had had enough attention up would come his beak and the serrated edge could prove very painful. He adored escaping from the waterfowl enclosure, escorting ladies to the loo and dutifully waiting for them.

Beppo, a blue and gold macaw, eventually earned the name "Beastly Beppo". He had a deep aversion to spectacles, especially mine. He would perch in wait and then dive-bomb me in a flurry of blue and gold feathers. I hated it. It was an awful experience, particularly if he took me by surprise. I learnt to become quite an adept Beppo-dodger.

Irrational dislikes often happen with birds. It then comes down to the simple ultimatum: "It's either me or the bird". We were given several birds which had crossed this line too often. On one occasion a woman told us: "It's either my husband or the bird" which did make me wonder what the husband was like.

A family named Edwards presented us with a Scarlet Macaw which we decided to call Edwina. She had very definite likes and dislikes. She would terrorise the public at times and had to be retrieved on a broom handle and banished to her aviary when she began to act up.

She had an unpleasant way of sidling up with a very quiet, discreet "hallo" and then attack or frighten anyone nearby. One day she managed to slip through the cafeteria into the kitchen. Three of the staff were so frightened they dived into the larder and locked the door. It took some time to rescue them.

But Edwina had a kinder side. She would always come over to our bungalow and knock on the French windows if either of us had been away or ill. One day she did her usual knocking and Don and I couldn't understand her message. We had not been away or ill. Edwina died that night. It was her way of saying goodbye.

I hand-reared Patty the greater Patagonian Conure. She was a dear and would fly to my hand when called. Patty gave a great deal of pleasure to the thousands of people who watched her in flight.

Oscar, the Oyster Catcher, does too - and causes great amusement these days. Some visitors come specially to see him. He rules the lawn opposite my bungalow very firmly indeed, busily twittering and making a raucous noise. With blazing eyes, he charges anything and anyone in his way. Not long ago he was brought back to the lawn after spending the winter in an aviary and looked thoroughly disreputable. Oscar had not previously met the gang of peafowl who had encroached on what he regarded as his territory. They were horrified at the sight of him and the aversion was mutual. The peafowl began to chase poor Oscar and he ended up having a nervous breakdown under a fir tree, which is where I eventually found him. It all ended happily. The birds made a sort of truce and now tolerate each other at a distance.

Besides these avian characters, we made many human friends in the world of zoos and bird centres.

One was the very distinguished wildfowl expert, artist and broadcaster Peter Scott. Don had first met him in 1951 not long after Peter opened the Wildfowl Trust at Slimbridge in Gloucestershire. They became acquainted through the Avicultural Society and Don used to go down from Dudley Zoo to see him. He had the greatest affection and respect for this brilliant, versatile man.

Don took me to meet the Scotts shortly after we were married. It was a great privilege to have known Peter and his wife Phil. A few years after our opening, Peter was the guest of honour at a champagne party we organised to raise money for the World Wildlife Fund. Peter was then the chairman of the fund. Nearly 200 people attended and we raised £50. Johnny Morris, who has been such a friend to us over the years, was there too.

A guest who found it a very nostalgic event was Captain Arthur Batten-Pooll VC. Rode Manor had belonged to his parents and he had spent his boyhood there.

I well remember Peter's kindness to Don the last time we

met when the Scots invited us to tea at their home. The living room had a huge picture window facing out on a lake which was always a mass of water birds. Don was already suffering from sight problems and Peter was so patient trying to help him see some of the huge number of birds outside. Peter died soon afterwards.

I am afraid that many of our friends have now gone to the Heavenly Zoo. Gerry Kirkham, who collected birds for us in East Africa and who was our host in Kenya, is one. Others include George Motterhead - "Motty" from Chester Zoo, who had been so courteous to me shortly after I first met Don - and Reginald Greed, the director of Bristol Zoo. George Cansdale of London Zoo has joined them and, of course, Don himself. It does rather make me wonder who is charge of each section of the celestial collection!

Another dear friend was our old postman Dick. I opened our front door one morning to find the mail scattered all over the place and that wicked magpie Freddie Maggs having a wonderful time sorting out envelopes on the edge of the pond. I left a note for Dick asking him in future to put the post under the doormat. A few days later I opened the door to find a pile of parcels several feet high topped by the doormat with a note pinned halfway up: "Your post is under the mat." The parcels, I discovered, were from my son Keith who was coming home from service with the Army in Borneo and had sent some of his belongings by post.

Another time, Dick left a large stone wallaby at the top of our drive.

Dick had a wicked sense of humour. We once received a card saying: "The Pet Refuge Assistance Bureau announces selecting your home for an animal shelter. Lodging and sand box facilities for 40 pigeons, 38 dogs, 113 cats, 17 guinea pigs and 6 kangaroos should be ready for occupancy by 2 a.m. Sunday. Special diet menus, exercise instructions and bath schedules will be provided for each animal."

I was absolutely horrified and rang the local Press, trying to make some sense of it all.

Then I suspected that it had to be a practical joke. At first I thought it must have been the work of a solicitor friend who loved these sorts of pranks.

And then I discovered the truth. Our stove had broken down so I popped down to the local pub to buy a take-away. It was there I spotted a card very like ours. It was an alert for a special fire practice which would involve umpteen firemen clambering on the pub's roof at the unearthly hour of 3 a.m.

The landlord's wife, Margaret Payne, laughed and told me it was Dick's handiwork. He had played a similar trick on a local farmer with a card saying that he had won a rhino in a raffle. I had to ring the Press again to tell them what had really happened and all three cards were featured in the local newspapers. I'm not sure how pleased Dick was about that.

Don couldn't resist getting his own back on Dick. He found a log and fashioned some eyes which I painted. We let it float in the pond outside the bungalow and it looked just like the head of a crocodile. We put a note outside our home warning: "Beware of the crocodile." That was for Dick.

As things turned out, this particular joke was taken rather seriously by visitors. They stood by the edge of our pond watching the log for ages, hoping to see the "crocodile" move. Sometimes we would slip in among the groups on the poolside to listen to their remarks. "There, I saw it move." "No, it's dead."

During the big Pick-Up of 1974, when I helped a Radio Bristol collection of lollysticks and metal cantops, Dick would stagger in each day with all the parcels plus a funny note which I somehow had to cap. How I wish now that I had kept all of them! We so enjoyed Dick and his sense of fun and life was never the same after he retired. He was a real breath of fresh air . . . a true original.

George Cansdale and Phil Drabble brought us many happy days. I did an interview with Phil for the BBC Radio Bristol programme "Calico Pie". He had just done the first "One Man And His Dog" programme and was a little concerned that it would only last for one series, although he had so enjoyed making the show. In fact it proved to be a marvellous idea for TV and the show has, deservedly, lasted for many years.

Don first met the delightful naturalist Gerald Durrell - zoo keeper and author of such classic books as "My Family And Other Animals" and "A Zoo In My Luggage" - during his Dudley Zoo days.

Gerry had been collecting animals in South America with Ken Smith and they sailed back to Liverpool docks. Zoo directors throughout the country received the message from Gerry that he had a lot of birds and animals on board and was prepared to sell some of them.

Don went to meet him. Gerry was suffering from malaria. He told Don and the others: "Look, chaps, I feel awful. Go and help yourselves to what you want and I'll send you the bill later."

Don had his eyes on a capuchin monkey, a Sun Bittern and two macaws. Our friend Motty from Chester Zoo wanted both macaws too. One had a club foot. They agreed to take one apiece and toss a coin for their individual bird. Don lost and had to take the club footed macaw, but it was still a delightful acquisition.

We visited Gerry at his home in Jersey several times and much admired the zoo he had created there from scratch.

Johnny Morris has been a great supporter over the years and I have written about our work together on TV in a separate chapter. When we held our 21st birthday celebrations at Rode in 1983, Johnny and another of our TV friends Terry Nutkins came along as guests of honour. The party was our way of thanking everyone who had helped us over the years . . . by then more than 2 million people had visited the gardens.

Other friends included Molly Badham and Nat Evans of Twycross Zoo. Don had known them before we met and married and I felt very privileged getting to know them well over the years. We were both very impressed with their success in breeding chimps and other animals. It was always a pleasure to stay with them and enjoy a cup of tea with a tame chimp.

Don, Molly and Nat appeared on TV together with a chimp called Melody. Melody had been presented to Dudley Zoo as a baby during Don's years there and was an old friend. She was so tame that Don used to take her with him to functions and she always behaved impeccably. Not, unfortunately, on this occasion.

Melody somehow got loose in the middle of the broadcast - television was live in those days - and created absolute mayhem in the studio. I'm sorry to say that I never saw the show, but it must have been hilarious to watch. Melody, I am glad to say, is still alive and well.

One day Don was talking to a visitor to the bird gardens who told him that his father had known him in his Keston days and had bought birds from him. He proved to be Lord Hugh Russell and Don remembered his father, the Duke of Bedford, very well indeed. We soon became good friends. When Hugh and his wife moved from their home in Wylie in Wiltshire they presented us with a macaw and several other birds.

The fun goes on. Only the other day we spoofed the nation with a story that, in conjunction with the University of Bath, we had created a live Dodo chick by cloning DNA from a 200-year-old femur.

I pretended that, on a trip to Zanzibar, I had bought a bone of a Dodo. It was, actually, a photograph of a Dodo skeleton. So between the University, our Head of

Conservation Mike Curzon and Radio Bristol, we concocted a Jurassic Park-style story that the world's most famous extinct bird lived again.

By coincidence the cricketer and former England captain, now sports commentator, David Gower visited us just as we were adding the finishing touches to the story. David was taking part in a charity walk in aid of the Macmillan Nurses and he was delighted to join in our hoax and be interviewed on radio to lend authenticity.

In some ways there was a serious side to our spoof. After all, if Don and Rode Bird Gardens had been in existence before the last Dodo died, we would almost certainly have found room for some as part of our continuing work alongside other zoos to protect endangered species.

The day the story was released? April 1, of course, and Johnny Morris did his bit for us by being interviewed on his own cloning success of a tadpole which turned into a whale!

Other like-minded friends included Roger and Mary Cawley from the safari park at Longleat, not far from Rode. We were having tea with them one day when in strode a baby hippo which they had bred.

For non-zoo or animal people, it must seem a crazy world that we live in . . .

Brolly . . . TV superstar.

CHAPTER NINE

TV, Radio ... and Brolly the Superstar

Wildlife - particularly exotic wildlife - has always been a very popular subject for the viewing and listening public. By the 1950s pioneers like David Attenborough, Jacques Cousteau and Armand and Michaela Denis had become stars thanks to their fascinating programmes.

TV and radio were keen to capitalise on this huge audience . . . and zoos were equally keen to help them and win invaluable publicity.

Don had found at Dudley that publicity was an essential ingredient for success. He couldn't care less how or why the name of Dudley Zoo was mentioned - as long as it was. Any publicity, he insisted, was good publicity.

In his Dudley days Don thought nothing of taking animals all the way to the BBC's studios in Lime Grove if there was a publicity opportunity. It was there he first met George Cansdale of London Zoo. George proved to be both a friend and a very useful contact.

When a programme was being planned, George would often ring Don and ask: "Any chance of a tame cheetah or a wolf cub?" Don always did his best to find appropriate animals for the broadcast and he and his head keeper would make the journey down to London with anything ranging from a small bird to a fully-grown camel.

Phil Drabble often came to Dudley for both TV and radio broadcasts and "News From The Zoos" with James Fisher was filmed there several times.

Any animal or bird which did something slightly unusual could hit the headlines and this publicity was almost always reflected in an increase in the number of visitors.

TV was very different in those days. Most of the shows - even TV drama - went out live and this posed problems of their own. Although the shows were always rehearsed, anything could happen once the cameras began filming for real.

News of a new bird garden opening in Somerset caused quite a stir as there was only the Wildfowl Trust at Slimbridge and Birdland at Bourton-on-the-Water in the Cotswolds at the time.

Not long before we opened, the very popular children's programme "Animal Magic" presented by Johnny Morris had begun. Don often took part, joining Johnny and Tony Soper. I well remember the time that Don, my son Andrew and I went to the TV studios for one edition. We brought with us Mr and Mrs Honey, our very tame Moluccan Cockatoos from Indonesia.

Tony Soper was doing a commentary on Alaska when we were led into the studio. Suddenly both birds began to shriek loudly, as only Moluccan Cockatoos can, and we were hurried out as quickly as possible, everyone praying that the viewers hadn't noticed such strange noises as they watched film of the snowy wastes.

On another occasion during rehearsals for "Animal Magic" I was asked to let a macaw fly freely in the studio. It missed Johnny's head by inches - an unrepeatable sequence in the days of live TV.

You had to keep your wits about you. During another broadcast when Johnny, the marine artist Keith Shackleton and Don were lined up and waiting for a chat on air, Don had become rather bored with the delays. Down came the message from the gallery: "Tell Don to stop looking as if he's waiting for a bus!"

Over the 21 year run of "Animal Magic", our bird garden

was featured many times. I well remember Johnny - by then a great friend and supporter - doing his commentary down by the lake alongside the flamingos and declaring that they had their knees back to front.

Shortly before we opened the newscaster Guy Thomas from TWW, the Bristol-based West Country independent TV channel, came to film our newly-arrived penguins. They had to be hand-fed with sprats . . . a very messy, smelly business. We suggested that Guy should have a go. I don't think he ever really forgave us for getting so fishy.

In the late 1970s I took a kinkajou called Kinky to "Animal Magic" several times but he became rather difficult and started to bite first me and then Johnny. So I suggested to the producer that he might instead like me to bring Brolly, the young Umbrella Cockatoo which we had bred.

Brolly had been thrown out of his nest by his parents and taken over by one of our keepers for a couple of days. He was then handed on to me to see what I could do with him.

Brolly was just over six months old when he made his debut on "Animal Magic" and he quickly became a star. He appeared regularly for three years and became very studio-wise. If he was asked to do something special like preening or talking, he would wait until the shot was set up and clamp his beak shut and glare with his beady little eyes as if to say: "Now what are you going to do?" The moment the cameras were off he was once again busily preening and chattering away.

When the hugely successful science fiction film ET was released, "Animal Magic" decided to do an ET special. Two walls were covered with shelves carrying ET models. Brolly was asked to go on to one of the shelves and for the first few minutes he had wonderful time, tipping the ETs off the shelf or trying to dismantle them. He eventually came to a very large ET with the most enormous eyes. Brolly started screaming blue murder. Johnny arrived in the studio to find bedlam.

Anything could happen on that show. One time I was in the studio with Brolly on my knee, both of us watching the magician Paul Daniels doing a trick with Rocky the penguin which was going to "transform" into a pygmy hippo. Unfortunately the hippo took umbrage at the wrong moment and began to run around the studio, frantically being chased by its keeper Roger Cawley from the Longleat Safari Park.

In 1983 the producer decided to film "Animal Magic" with Brolly taking tea with Gemini, Terry Nutkins' sea lion, at Terry's home in Skye. It was quite a business getting all the way from Rode to Inverness so the BBC arranged for us to fly up there. I arrived at Heathrow with Brolly in a small box, a cage and my hand luggage and I couldn't find a trolley or anyone to help me as I struggled with all these items.

When we finally got on the plane, the captain announced: "Welcome aboard, ladies and gentlemen, for your flight to Aberdeen". All the passengers began to shout: "No, not Aberdeen - Inverness!" I was very worried until we were reassured that the flight was to Inverness after all.

We were met at the airport but Brolly's box was suspiciously quiet. When I opened it in the car, a beady little eye looked up at me and he said: "You all right? Brolly's all right." Dear Brolly.

The hotel where we stayed was not far from the Highland Wildlife Park where we would be filming some sequences. It was a rather shabby place - luckily, as things turned out. Although there was a catch on Brolly's cage, I had no idea that he had discovered how to open it and get out. When I came back from dinner, there was Brolly trying to eat a rather tatty wardrobe. I brought a padlock the next morning.

Our first day was spent touring the Highland Wildlife

Don and I with Sara Kennedy when the Animal Roadshow came to Rode in 1988 . . . Sara was very much on edge because she was extremely nervous with birds!

On the same day, with Desmond Morris, her co-presenter. The author of The Naked Ape was fascinated by our bird gardens and returned for a fuller, private visit later.

Park before we travelled to Terry's home on Skye. The BBC had laid on an old, open-topped Morris Minor with a perch for Brolly on the seat beside the driver - Johnny Morris, naturally. The camera crew filmed Johnny and Brolly as the car drove through the lovely countryside. I, meanwhile, was squeezed down in the back of the car hidden by an anorak in case Brolly became fed up. In fact Brolly was as good as gold. He sat on his perch admiring the view and chatting to Johnny.

We all stayed at Terry's house. Poor Brolly had a dreadful shock on the second night. I was having a bath when I heard him screaming blue murder. Still dripping wet, I ran out to find that Terry's white cat had walked into my bedroom. Terry whisked the cat away and I managed to calm Brolly down.

We went up a very tortuous, winding mountain road and were filmed from a helicopter. Brolly, as professional as ever, never turned a feather even when Johnny had to reverse down for a second "take". The tea party, however, was a bit of disaster. The table-cloth had been set on a particularly boggy bit of turf. Johnny's language turned blue and Brolly didn't think much of the affair either. He wandered into some bushes and Gemini sat on the cake. It could not be shot again.

Not everyone enjoyed that trip but I certainly did. The weather was kind, we had a lot of fun and I found Skye a magical place. We were taken home to Rode by car. It was a long, long journey but Brolly was unperturbed, sitting contentedly on my knee.

When "Animal Magic" celebrated its 21st birthday, Johnny went up to London to be interviewed on TV and Brolly was asked to go along too. I travelled by train with Brolly in his cage covered in black polythene on the seat beside me. I got some very odd looks from fellow passengers when he kept saying: "Are you all right? Brolly's all right."

Brolly thoroughly enjoyed the broadcast. There was a large, appreciative audience and he ended up sitting on Gloria Hunniford's head and laughing. Whenever Brolly liked an audience he would laugh and that made them laugh too.

Brolly also appeared on "Gardener's World" with the sadly-missed Geoff Hamilton. Geoff was filming a sequence in Bristol under a fig tree which grew by a sewer on the river in Bristol and he wanted to explain how seeds of plants are passed through birds.

Once we reached the riverside in Bristol, Geoff had to clamber down a very steep bank and stand on the sewer with his back to the fig tree. Brolly was taken down to him and put on his arm. Geoff began his piece to camera. The crew were on the opposite bank.

Brolly thought the whole thing was great fun and climbed off Geoff's hand and up into the fig tree. This happened several times until one "take" when Brolly climbed down Geoff's leg, around the back and then reappeared around his other leg, looking up to listen to what Geoff was saying.

When the filming was finally over I was afraid Geoff might be furious about the whole thing. But he climbed up the bank, rather dishevelled and with a blood-stained hand from Brolly's claws, and was roaring with laughter. It certainly made a very different "Gardener's World".

That incident came back to me when I heard the very sad news that Geoff had died suddenly of a heart attack while he was on a bicycle ride for charity. I shall treasure my memory of that afternoon in Bristol and, like so many people across the country, very much miss Geoff's amiable presence on TV.

Brolly is now 18 years old - still young for a cockatoo - and he has his wife Boulan, but so far we have not had any little parasols.

Don enjoyed appearing on TV and was always at ease.

He did a weekly slot on HTV with Bruce Hockin in which he shared his huge knowledge of keeping birds as pets. With Bruce's help and without Don knowing, I made recordings of all the sayings of our Hill Mynahs. The sound engineers at HTV cut and arranged them and played them during the broadcast. Don got a terrific surprise!

We did - and still do - many radio broadcasts too. Soon after Radio Bristol started I was invited to take some birds to the studio for the children's programme "Calico Pie". I chose an Indian Hill Mynah called Sam which had recently been presented to us. The programme lasted for an hour

"SAM"

"Sam"

and Sam talked on and off throughout.

We often appeared on "Calico Pie". Sam loved it and showed amazing timing. If someone was a bit nervous on air and had just finished speaking, there would be a "Well, well, well" from Sam, making everyone laugh.

Spike and Ed, the show's two resident singers, wrote a song called "Well, well, well" which was recorded and sold in aid of what was then known as the World Wildlife Fund.

Sam helped publicise the Radio Bristol Big-Pick Up charity collection competitions, which I briefly mentioned on page 90. Children collected metal cantop rings and wooden lollipop sticks. The metal was sold to help the WWF and the sticks put in bundles as firewood for the old people in our village.

Sam caused chaos one day when we were filming "Animal Magic". He did a wonderful imitation of a cat miaowing. The producer heard and shouted to his staff: "Get that bloody cat out of the studio - there's a bird coming in for the show".

Sometimes we are asked for the impossible. One TV producer asked if we could bring along a peacock to the studio and make it spread its tail on cue. Yes, really! We tried to explain as politely as possible that peacocks don't display to order.

We are often requested to bring along birds to open fetes and village events. One day someone rang me and asked if we had an elephant - of all things - for her fete. Another time, a lady rang to ask me if one of our tame penguins would walk down the aisle for her daughter's wedding!

We are also asked to give talks with slides. Sometimes I take Brolly along. He and I went to a Probus meeting in a pub one morning. Before we started, I was told not to be upset if some of the audience fell asleep. Sure enough, one elderly gentleman was soon snoring quietly. Brolly sat at the top of his cage and watched the old man. He thought it was a splendid idea and when I looked around at him he had his eyes firmly shut too. I certainly had to keep talking that day with no help from my sleeping "assistant".

I was invited to give a talk with Brolly to a popular Women's Institute group. I was preparing the flowers for my mother's funeral the following day and the talk was in the evening.

But just before I left Rode, we discovered we had no hot water. The electrician checked our heating system and said we needed to call in a plumber, so we asked one to come as quickly as possible.

Brolly and I set off for my talk but when we arrived I was

told that a rather important member of the WI had not arrived and that I shouldn't start my talk until she was with us. At last she arrived in a great flurry, apologising and explaining that something had happened at home to delay her.

It reminded me immediately of an amusing parrot story I had once heard. I couldn't help telling it to the meeting. It goes like this:

A lady kept a particularly talkative parrot. One day she had a problem with some taps in her house and sent for a plumber. Then she suddenly began to suffer severe toothache and went for an emergency appointment with her dentist.

Meanwhile, the plumber arrived.

He knocked at the door. A voice called out:

"Who is it?"

"It's the plumber."

"Who is it?"

"It's the plumber."

"WHO IS IT?"

"IT'S THE PLUMBER" the wretched man shouted and promptly dropped dead on the doormat.

The poor woman returned after her dental appointment, looked at the mat and said: "Oh dear, who is it?"

A loud voice from indoors replied: "It's the plumber".

I couldn't help thinking about Don at home, going to the front door to answer the doorbell and asking: "Yes, who is it?"

"IT'S THE PLUMBER".

The audience loved the story and everyone had a good laugh including Brolly. The more he laughed, the more they laughed. I was spared having to give too long a speech since Brolly kept them so amused.

CHAPTER TEN

Our Travels

Living "over the shop" has its advantages. However you do need to get away at times, particularly if you are running such a busy enterprise as ours.

We hardly left Rode at all in the first few years. We had far too much to do as we built up the business and confronted the inevitable problems which arose as our collection grew and the bird gardens became better known. We managed a couple of very short holidays in Cornwall where we were lucky enough to meet Derek and Jeannie Tangye who had just started their bulb farm.

But Don wasn't happy at the prospect of leaving Rode for long. He said we couldn't spare the time to go gallivanting about. We needed to concentrate all our energies on getting Rode running smoothly, he insisted.

I had loved travel since my childhood and I was lucky enough to go off on my own to Cyprus in 1966. I adored it. You could visit both the Greek and Turkish areas in those days. I had a very useful introduction to friends-of-friends who worked in the wine business there. I was taken to many fascinating places and even invited to a Greek wedding. I think it was the Cyprus brandy which saved me from any tummy upset . . . the wedding lasted for ages!

Don was - and always remained - uneasy whenever the subject of travel cropped up. But we had so many friends in the wildlife world who had visited or lived in the far-off countries where our Rode birds originated that I became more and more determined to take Don overseas to experience wildlife in its natural setting. Don already knew an enormous amount about keeping birds in alien habitats. I was convinced it was time he saw the real world of his birds.

Not long after I returned from Cyprus, our good friend Gerry Kirkham insisted that we should join him in East Africa. Gerry had lived in Ireland until his wife died and then moved to Kenya.

Don was very unsure about the venture. My mother came to see us off at Gatwick, quite a small airport in those days. Just before we said goodbye she glanced at Don's glum face and told me: "I'm glad it's you going with him and not me."

Whenever we were about to set out on a journey and were waiting at an airport Don always had this pained expression which said: "I just want to go home." He hated flying and when the flight was announced he would vanish to the loo and there would be a panic to find him.

I loved that first flight to Africa. We flew past the Alps and I recognised many peaks which I seen or even climbed in my mountaineering days. Crossing the desert world with its sand dunes and sand waves was fascinating and I was sorry when darkness came and I could no longer see the landscape below.

When the plane made a scheduled landing at Entebbe, we went to the bar in the transit lounge and bumped into Don's old friend, the naturalist and wildlife collector Carr Hartley. Don had known him well in his Dudley days and Carr had supplied him with animals for the zoo. It was a great reunion and we were delighted when he told us he was also on his way to Nairobi. He joined us on the flight.

There was even more excitement when we at last arrived in Nairobi. Gerry was waiting for us - it is so much nicer being met by friends when you go to a strange country for the first time. We piled into Gerry's car and Don mentioned that he had spotted some antelopes on the runway perimeter when we stepped off the plane. Gerry told us they were Thomson's gazelles and that they often grazed there.

As we weaved our way from one side of the road to the other to avoid potholes, I asked Gerry which side of the road you drove on in Kenya. "Oh, the best" he said casually.

Gerry shared a house with his friends John and Jenny Pearson who worked for East African Airways and we stayed there. We met lots of their friends and had a very warm welcome.

Gerry and his friends Lucy and Ivan Bampton, who were to join us on our safari, had planned a wonderful three-week trip taking us up into Uganda, down into Tanzania and then across to the Kenyan coastal resort of Mombasa.

Once on the road, Don was thrilled with all the bird life. Because the East African countryside is open savannah and birds have to fly from point to point, they are very visible, unlike in forest country. Don had known many of these birds in captivity. Now, at last, he could see them at their best.

We saw a great deal that today's safari-goers would never be able to see. We went to the Budongo Forest in Uganda where we watched chimps in the early morning. Ivan, who was collecting butterflies in the forest, took us out during the day to set his traps. We returned at night and I could feel eyes everywhere and had that incredible sensation of being watched by all sorts of creatures, large and small.

A special boat took us to the foot of the Murchison Falls in Uganda. The area was densely populated with hippos and giant crocodiles up to 20 feet long. One or two of the hippos came towards us and swam under the boat.

Some of the best bird-watching is on the lakes in Kenya. We arrived at Lake Naivasha and Don counted 10 different species of water-bird in less than a minute. He was enchanted as he watched Goliath Herons, pygmy kingfishers, waders and ducks. Before we left Lake Naivasha he saw a small flock of Crowned Cranes in flight. They honked and flew away, their great wings spreading. It was a particularly special moment: Don had chosen the Crowned Crane as our emblem at Rode.

One of the most spectacular sights I have ever seen was at Lake Nakuru, the soda lake famous for its flamingoes. There were so many that they had to be counted by the acre.

We visited Mount Kenya and stayed at the Mount Kenya Safari Club, which was wonderful. Ivan caught some Golden Rain Sun birds and we later brought a pair back home with us.

One of the highlights was our stay at the Tsavao Kilaguni Lodge. Although it was out in the wild it was extremely comfortable with all the mod cons and was largely patronised by European and American tourists.

The dining room was like a huge, open-fronted veranda. We sat in subdued light above an artificial lake illuminated by headlights trained on the water and the surrounding area. It was a little like a stage set with wonderful animals as the actors.

They came to drink in ones and twos. First an elephant, and when he disappeared, a young rhino. The animals materialised out of the darkness, gave their performance and then vanished into the night. All the while bats flew across in front of us.

We flew into the Serengeti in a light aircraft. During the flight we passed the crater of the volcano Longhi which had erupted earlier. The pilot knew I wanted to take photographs so he banked the plane for a better view. As the plane turned, Lucy piped up: "Anyone want a banana?" Don groaned and slid lower in his seat, hiding his head in his hands. Don was petrified the plane would be sucked into the volcano. Later we made a trip to the Ngorongoro Crater where we camped on the rim with a fire at night and strange animal noises to lull us to sleep.

Flying in light aircraft in East Africa can be hair-raising if you don't know what to expect. Once the sun has warmed the plains, it creates up-currents and air pockets and your plane can plummet or rise a couple of hundred feet without any warning. Don said it was like being in a lift that had gone wrong.

On all our drives with Gerry - a terrifying driver - we saw a profusion of wildlife as well as birds. Elephants, lions, zebras, giraffes, wildebeest, leopards and much, much more.

Towards the end we travelled overland from Nairobi to Mombasa, 300 miles away. Over here and with our good roads, most of us would consider that a very long drive. Gerry, bless his heart, didn't leave until 12 noon and we took a picnic with us. East African picnic lunches were huge feasts. The journey along dirt roads was a nightmare. One area - the Kibwassi Drifts - was like driving on sheet ice. We skidded several times when Gerry was doing 70 mph. We had endless punctures and had to prop the car up with stones while running repairs were carried out.

We were all praying that we would reach Mombasa in one piece without a further puncture. I'm glad to say we made it. The Kenyan coast is particularly beautiful and we stayed in simple little huts, falling asleep to the sound of bush babies. The sea was almost at our door and the beaches were wonderful. We were taken by dug-out canoe to the reefs and allowed to wander around seeing the beautiful corals just below the surface.

On our way back from Mombasa we visited the Amboseli National Park. We had hoped to stay at the park Lodge but when we arrived we were told it was full so we had to press on to the exit. Don had become so fed up with Gerry's driving that he took over the wheel and did, I must admit, drive rather slowly. By the time we had reached the park gates the sun had set and the gates were locked and guarded.

The guard asked us: "What are you doing in the park?"

"Trying to get out."

"But you cannot get out. The gates are closed now."

"We'll have to stay here then."

"But you cannot stay in the park after dark."

"What shall we do, then? Shall we wait until we are eaten by lions?"

"Oh no, that is not allowed."

This mick-and-mack conversation went on for ages until he finally relented and unlocked the gates for us.

We returned to Rode full of our experiences. That first visit to East Africa had opened our eyes to a new world and it had given us an appetite for more.

Sure enough, we were back a year later.

This time we joined the naturalist John Alexander for a tented safari on the Tsavo river in Kenya, a fantastic opportunity for really close-up animal viewing. John proved an outstanding safari leader.

Don noticed how quickly the animals picked up our scent. One evening a herd of elephants was approaching us from upwind when, at quite a distance, they caught our scent and raised their trunks to sniff the air. They clearly didn't like what they smelt and turned around.

Following this incident, John said: "Come on, let's have a bit of fun. Let's see just how close we can get to them."

Two young bull elephants came down to the river so we silently followed John as he crept along among the undergrowth. Before long we were so close that we could even hear their tummies rumbling. Then, quite suddenly, they caught our scent, wheeled about and were gone.

My interest in photography had by now become intense. I had learnt on our previous trip that if I was going to take good safari pictures, I needed a long-distance lens, so I had brought with me my new Nikkon with a range of lenses including a 400/600. My photographic equipment was fairly heavy, but it proved well worth all the effort of lugging it around and I have my wonderful close-ups of these elephants to this day.

John's tented safari was beautifully organised. There was every possible comfort that you could want in the wild. Washing and laundry was done superbly every day and the food was excellent.

I received an unexpected invitation to join Carr Hartley on a rhino-catching expedition and Don encouraged me to stay on in Kenya to take part in this adventure. I decided to spend a few more days with John's safari while Don returned to Nairobi on his way back to Rode.

Don made very good use of his stop-over in Nairobi before catching his flight. Gerry had found him a young Secretary Bird which had fallen out of its nest and been hand-reared. He asked Don if he would like it for the bird gardens.

Gerry sent a cable to the tour operator. "Can passenger Risdon bring back a Secretary Bird instead of wife."

The paperwork was completed - in those days it was far easier to import exotic birds: today livestock has to be sent by freighter or put in a separate animal compartment if it is a passenger plane - and Don arrived at Nairobi Airport in the blazing sunshine with the Secretary Bird in its crate. The crate was left out in the open and Don became very concerned in case it caught sunstroke. He strode out onto the tarmac and calmly asked if it could be moved into the shade beneath one of the aircraft's wings.

Once aboard, the crate was placed at the end of a gang-way between the ladies and gents' lavatories. Don's fellow passengers were rather amused by their rather unusual travelling companion.

When he arrived at Heathrow, a Customs officer asked about the contents of the crate.

"It's a Secretary Bird. Would you like to see it?" Don asked.

"I'd love to. I've never seen one of them".

Don let him have a peep and the officer was so fascinated and intrigued by the unusual-looking bird that he waved

Don through at once and told him to get the bird to Rode as quickly as he could after its long journey.

This Secretary Bird was with us for 22 years. He was given a mate several times but, sadly, never managed to fertilise any eggs. Today we have another pair and have great hopes of at last breeding this species, a feat rarely achieved outside Africa.

Meanwhile I missed Don and, to make matters worse, had an upset tummy. I felt so wretched that John arranged to have me flown back to Nairobi by light aircraft. Back in Nairobi I was fine after a couple of days.

Carr, his wife Daphne and I had an awful job getting both ourselves and his equipment into Uganda and being granted the necessary permits, but at last we made it. There had been a lot of rhino poaching in the area and Carr was determined to see what he could catch and move to the National Parks for safety.

After the luxury of John Alexander's safari camp, this was very basic. It rained, I was bitten by every imaginable bug possible and the drives in the Land Rover were very, very rough. I'm afraid we never succeeded in catching any rhino but Carr did teach his Africans assistants how to catch a great variety of animals including giraffe, zebra and antelope by using a combination of tranquilliser darts and ropes.

We were the last vehicles to leave the Serengeti that season and we only just made it in time. We were badly hit by storms and passed lorries which had become completely bogged down in the mud.

My flight home was just as eventful. It was at the time when Idi Amin was starting to victimise Asians in Uganda. Many crossed into Kenya hoping to get tickets to fly to the United Kingdom. Air tickets to Britain were at a premium and I could have sold mine for many times its face value, but my conscience wouldn't allow me. Don needed me at home.

I stayed, as usual, with our friends John and Jenny Pearson in Nairobi and luckily Jenny was the stewardess in charge of First Class on my flight to London. It was chaos. Refugees were everywhere, trying to find places. Jenny escorted me to First Class and I had a wonderful journey home.

When we landed in Rome for re-fuelling, who, of all people, should I bump into but our friends Roger and Mary Cawley from Longleat Safari Park. They were on their way back from Africa too. What a small world it is.

Later that year Don and I attended the preview party for the Wildlife Artists - my painting of our Secretary Bird's head was one of the exhibits - and I was introduced to Jean.

We began chatting to Jean about our experiences in Africa and she was spellbound. She said she longed to go herself. In the end Jean and I agreed to make a trip the following year. She was already keen on photography and I assured her there would be no lack of subjects.

I contacted Carr in Kenya and he invited us to join him on one of his hunting sessions around Arusha. He suggested spending a week with him and then going off on our own.

The trip began very badly. There was an extremely boring drunk on the plane who kept everyone up all night. We journeyed to Arusha but since there wasn't time to join Carr that evening we stayed at a nearby hotel. It was a lovely place but we had no time to enjoy it. We had to leave at 3 a.m. to meet up with Carr armed with a packed meal which turned out be some stale sandwiches and Thermos flasks full of hot water. They'd forgotten to put in the tea. I began to think, Oh God, why did I ever say we would come on this outing?

Then, suddenly, the sun rose over the top of Mount Kilimanjaro and I knew why . . . it was worth every minute of discomfort. We arrived at the camp just before the start of a rhino-hunting expedition.

Those were very exciting days. Our transport included a helicopter from which vets would dart animals with tranquillisers. Our job was to follow in two open Land Rovers and a lorry to carry away any darted creatures.

Jean and I joined the second Land Rover. We roamed up and down, following the low sweeps of the helicopter. Suddenly they spotted a rhino, darted it and radioed us its location. We crouched as low as we could in the vehicle to avoid being scratched by the thorny bushes we charged through. We couldn't find the rhino. We three women, Jean, Carr's wife Daphne and I were left in a Land Rover while the men went off on foot to try to find it.

We tried to pass the time by joking, but it honestly wasn't in the least funny. We were very much aware that a woozy rhino might be staggering our way in a foul mood. They never managed to find the animal and we returned safely to camp. I'm sure the drug wore off and the rhino eventually managed to recover perfectly well.

The following day I was allowed to join the helicopter crew. I was very keen to photograph the area. Before long we had spotted and successfully darted a young rhino. We landed and went plunging off through the bush. I was snapping away like mad and have the whole sequence on film . . . the placing of a sunhat over its eyes to prevent sunburn, the loading into the lorry and the vet injecting an antidote - and jumping clear of the lorry as quickly as he could.

The rhino looked fine when I saw it in an enclosure not long afterwards.

By now I had a secret ambition to become a part-time professional wildlife photographer, combining it with my work at Rode. I dreamed of persuading Don to let me visit various parts of the world so that I could build up a really good portfolio and Here-and-There photographs, Here at Rode and There in the wild. It never really happened but I do have a wonderful collection of pictures and used to give illustrated talks using some of my slides.

Jean and I went through miles of film during those stupendous hunting days. Jean's equipment was a bit limited so the moment we arrived back in Nairobi, she rushed off to buy better lenses. We hired a car and set off into the wild. We took it in turns to drive, the other sitting in the back so that if there was anything worth photographing, we could stop the car and had our own windows through which we could shoot.

Jean has done a tremendous amount of photography in the wild since then, particularly of whales.

My next big trip with Don was to Australia. I was so pleased when I finally persuaded him to go. I longed to return to the country where I had been so happy as a girl. We decided to fly westwards rather than east so that Don could meet up with one of his war-time RAF pals. I stupidly broke my finger by slamming a safe door on my hand in our rush to get away from Rode and the air stewardesses on the plane made me a little splint out of plastic teaspoons.

Los Angeles made a convivial break but the flight from LA to Sydney was very long. We stopped to re-fuel at both Hawaii and Fiji and by the time we touched down in Sydney we were both exhausted. The immigration formalities seemed to take forever but, finally, we were safely in our taxi en route to our motel in Manly. I'd warned Don that the drive from the airport to Manly took us through some pretty grotty areas but it didn't help.

Don looked out at the decidedly unlovely view muttering: "Ugh, bloody Australia. What on earth did you want to bring us here for?"

Once we had checked into our motel, I arranged to see a local doctor immediately about my painful broken finger. We had to hang around in the waiting room for hours and, halfway through, Don said he'd had enough. He couldn't stay awake any longer and was going to walk back to the motel. I asked him to pick up some sandwiches on the way

The broken finger.

- we hadn't eaten since breakfast on the plane at 5 a.m.

At last I was seen and my finger treated. I walked the mile or so back to the motel and found both my sandwiches and a fast-asleep Don. I joined him. Our sleep was interrupted by a call from a friend of my parents in Canberra who wanted us to visit him so that he could show us the local Lyre Birds.

"I don't want to see Lyre Birds. I just want to go home" Don grunted. "And turn that bloody light out."

I couldn't get back to sleep. I felt miserable and hungry and my finger hurt. I went into the bathroom and thought: I'm 13,000 miles from home, it's raining, I'm hungry and I don't know what's going on. I tried to write some post-cards. It didn't help. The sandwiches had long since been eaten and I was ravenous.

I slipped on a pair of my son Andrew's old gym shoes which I had brought along for walking on coral and other sharp surfaces, I stepped out into the rain and began to search for somewhere to eat. I found just what I wanted - a friendly cafe where I had was given tomato soup and baked beans on toast. I can never forget the taste of that simple snack - it was one of the most welcome meals in my life.

It's amazing what a good night's sleep can do. Don woke up bright and happy. There was no more talk of wanting to go home.

We went to Sydney Zoo to meet its director, Sir Edward Holstram. He had been instrumental in founding the zoo and was a great bird man who had bred a lot of parrots and Australian parakeets.

At first Sir Edward was rather distant - I don't think he had any idea who we were. But as we sat rather uncomfortably in his sitting room, Don saw a large book on the table. The book was open at a page of illustrations of various parakeets. It was the turning point of our visit.

Don recognised what he thought was a Paradise Parakeet in one of the pictures. At the time it was considered to be a natural hybrid between the many coloured and hooded parakeets. Don had kept both at Keston and immediately said that he was convinced that the Paradise Parakeet was a hybrid.

The ornithologists of the day, on the other hand, insisted that the parakeet was a distinct species. Sir Edward knew better. He had bred several of these birds himself and Don's evaluation impressed him greatly. Don became his "blue-eyed boy" from then on and we got on like a house on fire.

Sir Edward was kind and generous to us in a way that only Australians can be when they take to you. He took us round Sydney Zoo and showed us his private collection of Albino Wallabies and introduced us to his various pet birds including a Slender Billed Cockatoo. They are marvellous talkers and this one could say practically anything in the broadest of Australian accents - things like "Do you want a job 'ere" and "Want a cup-of-tea?"

I had very much wanted to revisit my childhood home in the outback but there wasn't time. I still rather regret it, but perhaps it was best to leave it to my very happy memories.

We then moved on to my parents' friend in Canberra. He was keen to introduce us to a local bird expert and we were taken to a house where, just as we walked up to the front door, a voice cried out: "My God, it's Donald Risdon. You're the reason I came here!"

The bird expert turned out to be Terry Thomas. Terry used to live in Dudley, was a keen bird-fancier and had collected all of Don's bird-keeping books. He had been so inspired by Don's example that he had emigrated to Australia, determined to start his own zoo.

Terry took us to all the best local places to see wild birds. I'm sorry to say we never saw the elusive Lyre Bird in the wild but we did visit the woods where they lived and picked up some of their feathers.

Lyre Birds are very shy and difficult to see. But, as compensation, we watched a great many cockatoos.

We visited a local botanical garden which was a collection of purely Australian trees and plants and Don was particularly pleased when he spotted Australian Fairy Wrens. They are minute birds, about the same size as our native wrens but with a longer tail. They are vividly coloured in brilliant sapphire blue, scarlet and green.

Don enjoyed the company of many wild birds he had only known as caged or zoo specimens . . . a party of Pennant Parakeets squabbling over a nest and Gang-Gangs flying above us. Over here they are regarded as a great rarity but they were quite common in that part of Australia.

We sighted a fair number of Roseate Cockatoos or Galahs, not in those huge flocks you see in natural history films made in the dry, arid interior but it was still a pleasure, and we also saw Red Rumped Parakeets.

Don happened to mention to some local farmers that Galahs are very valuable back in the UK and, even in those days, fetched up to £500 apiece.

One farmer was absolutely astonished. "Over here we shoot the buggers. They're pests and we organise Galah shoots."

We learnt about the total ban on the import and export of wild birds and animals into Australia. It is a very understandable precaution when you think about the devastating effect on local wildlife from introductions to Australia like rabbits. It has meant that Australian zoos concentrate on indigenous species and we only saw a few lions and tigers which were clearly relics from the days before the ban.

We travelled to Brisbane and the Gold Coast where I had my first true Australian surf swim for a very long time. We visited the famous Currumbin bird sanctuary north of Brisbane where the keepers have taught the wild Rainbow Parakeets to fly in for sweetened bread and milk. In the wild, these birds live on nectar from flowers and the local birds have gradually developed a liking for this substitute feed.

Visitors were encouraged to take seats in rows. Then the old boy who organised the feeding came across with a bucket covered with a cloth. The birds swarmed all around him, trying to suck the nectar through the old dishcloth. Then each of the visitors was given a plateful of the feed. We just sat there and watched more and more of these gorgeously coloured birds flying in.

I shall never forget the blissful expression on Don's face as several of these birds sat on his arm and hat while he fed them. He was very moved by the experience. He said afterwards that it showed how harmony could exist between man and wild creatures, neither molesting the other.

We flew up to Mackay to visit the Great Barrier Reef. I had always wanted to go there and it was every bit as exhilarating and beautiful as I had hoped. We flew by helicopter to the beautiful island of South Moll for a week's stay.

It was paradise. There was every sort of facility from scuba diving to floodlit tennis . . . and wonderful walks. I was fascinated by how effectively the bright colours of the parrots and parakeets worked as a camouflage among the eucalyptus trees. Somehow their brilliant colours merged perfectly with the light and shade of the foliage and you really had to look long and hard to see the birds.

We took walks in the hills where Don saw what at first he mistook for vultures. In fact they were Flying Foxes (fruitbats). We found a hillside spot where we were on eye-level with the tree-tops and watched them fighting and squabbling, hanging by one hind leg with their wings wrapped around them like cloaks.

I went snorkelling . . . how can I ever forget the colours of that other world lying just below the surface.

We had to leave in a terrific hurry. After four months of drought, storm clouds were gathering and the helicopter pilot was keen to get us back to Mackay before the storm struck. We made it just in time and finished our Australian adventures with a wonderful time in Sydney thanks to the hospitality of our new friend Sir Edward Holstram.

When we parted near the airport, Sir Edward turned to Don and said: "Well, goodbye old boy and I hope to see you again one day". Sadly, he died only a few months later and we never did meet again.

I couldn't persuade Don to join me the first time I went to Trinidad and Tobago with my son Andrew. We had had an invitation from Dick Dean who was in charge of the Wildlife Trust in the south of Trinidad.

First we visited Tobago where I met the charming Chinese gentleman Mr Lou who kept the Bird of Paradise Inn on Tobago. Birds of Paradise had been introduced to Tobago from Borneo and New Guinea. Mr Lou was then well into his 70s and had a burning ambition to create a 10-acre walk-through aviary featuring these spectacular birds. We stayed at his rather run-down hotel and I couldn't help noticing that his birds were not kept as well as they should have been. I also noticed his Royal Suite - it was kept in the unlikely event of a member of the Royal Family coming to stay! But he was a charming man and he insisted that the next time I came, I should bring Don.

I then went to stay with Dick Dean and his wife in their house south of Trinidad. Dick took me up into the northern range of mountains and showed me the collection of water fowl they had built up. We were talking zoo business one day when I mentioned our labelling system at Rode and offered to prepare some for him. I heard only recently that my labels are still there.

A couple of months after I returned to Rode, I received a cable from Mr Lou who clearly was not going to take no for an answer about Don visiting Trinidad and Tobago. It simply said: "I have booked your room."

I had to do a fair bit of arm-twisting, describing the wonderful bird life I had seen. Don wasn't keen but at last he relented. He never regretted it. Don disliked hot weather and he found the temperature in Trinidad difficult to cope with. Tobago, a small island with cooling breezes from the surrounding sea, was much more to his liking. We did indeed stay with Mr Lou . . . and I was horrified when cockroaches came out of the peeling wallpaper! Don, on the other hand, was fascinated by these horrible creatures. Typically, he was interested in even these forms of wildlife and was much amused by my fright.

Despite the heat in Trinidad, Don found plenty to enjoy. The highlight for him was to watch huge flocks of scarlet ibis and white egrets flighting in each evening. Silhouetted against the bright blue tropical sky, the birds transformed the mangrove swamps when they landed, making it look as if the place had suddenly burst into red-and-white blossom.

We went to visit a friend's house where they had put out nectar feeders to attract humming birds. Don's knowledge of humming bird species was fairly limited but he was enchanted by the sight of these tiny birds coming in to feed, fighting and clicking their little beaks.

On my previous trip I had met a remarkable English lady bird-lover who was determined to do everything she could to encourage local bird life. The windows of her house were glassless and open to the elements so that birds could come in and out as they pleased.

Don and I went to stay in her very unusual household. As we lay in bed in the mornings, we could hear the approach of the Cocrico, a species of Guan. We could hear them in the distance, a little like a far-off football crowd which was coming nearer all the time. They poured in through our bedroom window and then out through another.

There were other trips. To Florida to see the Parrot Jungle which I had visited years before, to the Grand Canyon and to the Nile.

But the best, almost inevitably, was the first . . . . that heavenly visit to East Africa. Don, who had learnt his bird lore with captive creatures, had seen the true nature of exotic birds in their home setting at last. I was so glad that I had been able to give him a new, far fuller perspective on the creatures he loved so much.

CHAPTER ELEVEN

Gardening at Rode

We have some wonderful sepia photographs of the pleasure grounds at Rode Manor taken before World War One. The gardens must have been absolutely magnificent.

The photographs show fine terraces, tall, lovingly clipped yew hedges and the decorative lake, complete with a little punt. The grounds were a very fine example of formal English country house gardening and must have required a great deal of labour to keep them looking their best.

The only worthwhile legacy we inherited was a lovely back-cloth of good trees, dating from those days of splendour when, doubtless, a small army toiled away keeping everything just so. We had no intention of even attempting to restore the grounds to their original condition, but we did want to make our 17 acres as attractive as possible by creating some semi-formal areas and tidying up the woodlands.

For the first few years I was so tied up with catering and my many other chores that I did not have time to do much more than to bring in the odd plants and plant them in a haphazard sort of way. That did not stop me from longing for the time when visitors would say: "Oooh! How lovely" when they arrived.

In the early days when we used to visit other bird gardens and zoos, we often wondered why the owners did not add some colour by planting flower beds. However this was certainly not the case at both Bristol and Chester Zoos. They are great rivals for the accolade of the best zoo garden in the country. When we returned from one of these visits Don

Brolly with Clematis "Lasurstern", part of my clematis collection at Rode.

would say that although we had lovely trees, we really did need more colour.

Don always said that my first effort at gardening at Rode was to grow mushrooms in a bucket. Very good they were too, but that was hardly a help in making the gardens more attractive.

The entrance beds, which are our shop window, needed impact. This was easiest in spring-time, with masses of bulbs and plenty of colourful blossom.

When we extended our bungalow in 1970, we decided to establish a garden around our home. So we went to see our friend Gerald Thompson at Oldfield Nurseries in nearby Norton St Philip. Gerald suggested that among other plants, we should try some clematis. He warmly recommended "Mrs Cholmondley" as a pleasant lady. Sadly, "Mrs Cholmondley" did not approve of her new home at first, but I now grow her successfully in another part of the garden. She is a good but rather blowsy lady.

Don loved our bungalow garden and its pond and he introduced several ornamental ducks. His favourites were Mandarins and Carolinas which we rear each year. They stay on our pond until they become full-winged and fly down to the lake. They always fly back to the pond to be fed.

I have my very own natural lawn-mowers - Red-breasted Geese which keep the grass down as well as looking attractive. A little fountain plays on the pond and keeps the water fresh. This is very necessary if you want to keep water fowl. Don's advice on the subject was to ensure that you have a good water supply and good drainage so that you can empty and re-fill the pond easily and it does not become messy.

We also have a pond in our back garden where we grow water lilies. Don was particularly fond of them - he often spoke of the serene and calm feeling they gave him.

I began to take a keener interest in gardening and joined

the Rode Horticultural Society when it was formed in 1975. In 1983 our society had a day trip to Treasures of Tenbury and Burford House. John Treasure began this venture and had a wonderful collection of clematis. I came back with several plants including Hagley Hybrid, which still looks lovely in the back garden. The following year I took Don to see the garden and to meet John Treasure, who became a good friend. We brought back more plants and I decided to become a member of the International Clematis Society.

In 1985 I spent two days at the society's big meeting in London. I was very impressed by the variety of clematis plants and with the enthusiasm of Raymond Evison, the then chairman. I thought it might be an idea for us to start a collection of our own at Rode. Before long we had more than 100 plants in the collection and that figure has grown to nearly 200. Some 170 can be seen by our visitors and the rest are in my own private garden.

We had a long, cold, wet and gloomy winter in 1995/6 but the following summer produced one of the best years for clematis that I can remember. Some plants that had not flowered for a long time suddenly revived. One tip that I learned years ago was never to dig up a plant, even if you thought it was dead - it may come back after 2 or 3 years.

Another very useful tip is to keep plants well labelled. This can be difficult when selling them because the labels tend to get handled quite a bit, but it is worth the effort.

The British Clematis Society was formed in 1991 and has flourished. I was appointed treasurer and membership secretary and we started to go to shows and held annual weekends at Rode I have made many good friends through the society and now have a collection of more than 20 herbaceous clematis.

When I am gardening at Rode I need to be featherbrained - to be aware that birds will be at work all around me. Birds make very bad gardeners, especially peafowl. Here we also have problems with rabbits and squirrels. It is a never-ending battle against some creature or another. One time we had a dreadful gang of seven peafowl who belonged to a neighbour. They refused to stay at home and did a great deal of damage here. I'm glad to say they have since gone.

When I plant pots or containers with bulbs, I stick canes capped with upturned hanging baskets into the soil. It is the only way to stop the peafowl sitting on top of them and ruining the growing plants.

During the season I used to sell clematis at Rode. Over the years here and at shows I have had some very funny remarks:

"Why did my clematis die?"

"Did you water and feed it?"

"Oh, I didn't know I had to water it."

or

"What is my plant?"

"Does it have a label?"

"No, my husband died before he could label it."

"What colour is it and when does it flower?"

"Oh, it's pinkish, blue-ish, mauve".

One man I met at a show asked me to identify his plant. He couldn't remember its colour, when it flowered or the appearance of the leaves. In exasperation I said: "Why don't you bring me a flower and some leaves to the bird gardens and I'll see if I can help you." He never came.

It is interesting to see how popular clematis have become in the last few years and how busy any clematis stand is at the big shows like Chelsea, Hampton Court or Malvern.

One of the Society's first stands was at the Malvern Spring Show. It was quite a baptism. During the three days of the show the questions came quick and fast. "Where can we buy clematis? Can we buy those?" ("Those", in fact, were artificial clematis flowers!) They gave an excellent effect and were smelt and touched; one man said they were much better blooms than his at home. "Can I grow my clematis over a corrugated-iron roof or will it get too hot?"

It was great fun but very hard work. You had to smile through the unending questions, have patience, cope with all the weather conditions and not be like some stands where we saw the helpers sitting arms folded, grim-faced, with the look "I wanna be home an hour ago - so don't ask me any questions." I'm glad to say no BSC member was like that!

I wrote an article about our experiences during those three days at Malvern and was thrilled when Jim Fisk, Britain's best known clematis expert, used an extract from my article in his book **"Clematis: The Queen Of Climbers"** **by Jim Fisk (Cassell).**

The British Clematis Society has recently produced at least five fact sheets on planting, pruning, growing from seed and container growing. Often people say they have no room left in the garden - well, I can always find a corner for one more clematis.

Happily for me, clematis are reasonably bird-proof. The only exception is when a duck or peahen decides that the base of one of my plants is an ideal place to nest and I have to leave them undisturbed until the eggs hatch.

Bulb planting can be an awful chore but I have always enjoyed it because there is always hope - hope of a beautiful spring with the smell of hyacinths, narcissi and daffodils. People do not always realise that some of the clematis plants have a lovely scent.

It isn't just the peafowl which can be very unhelpful towards my gardening efforts. Sarus and Crowned cranes were perfect pests when at liberty in this way, always digging up plants or bulbs as soon as I put them in.

I have several favourite plants like the hardy (cranesbill) geraniums. I try to add to my collection each year. I tend to keep my pelargoniums and fuchsias protected over the winter. Years ago a pelargonium expert came to speak to our Horticultural Society in Rode. He had one particular plant

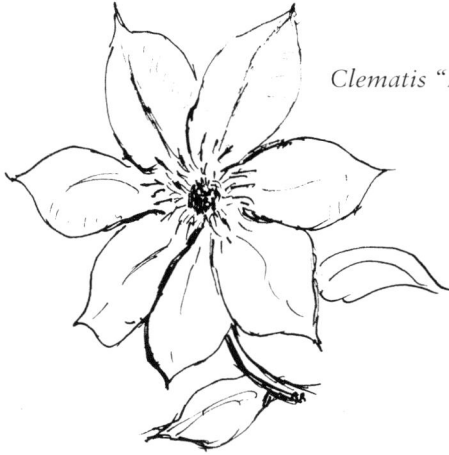

Clematis "Dawn"

Clematis "Miss Crawshay"

Clematis florida seibodii

which I greatly admired. It was a Royal Maroon. I told him how lovely I thought it was and he presented it to me. I have several of these plants to this day.

Years ago I was advised to dig up my fuchsias in the autumn, cut them down, put them individually into plastic bags with soil around each one and tie the bag loosely at the top. They are over-wintered in a greenhouse which is only heated in intensely cold weather. Over the years I have kept many plants this way and they are used in containers, hanging baskets and urns.

I do about 20 hanging baskets every year to go around the entrance and catering veranda area. Hanging baskets and container gardening have become very popular across the country, making everywhere look far more attractive.

The beautiful mature woodlands at our bird garden are one of our finest natural features. We started a Tree Trail here which was of great interest to our visitors because of the many fine trees planted here in the 19th century.

We added a further attraction in Easter 1988 when we opened the Rode Woodland Railway. This 7 1/4 inch gauge railway has brought pleasure to countless thousands of visitors and gives them a wonderful guided tour through the grounds. After leaving Woodland Central Station this scenic line weaves its way across lawns, past the pheasant enclosures and through wooded areas on its route of half a mile. The birds have got so used to the railway that they sometimes stand on the line and refuse to move, much to the amusement of passengers.

Because the line has steep gradients and sharp curves, one-third full size narrow gauge steam locomotives are used which are replicas of full sized engines. They are coal fired and operate just like their larger sisters.

The railway is run by steam enthusiasts who build their own locomotives and coaches and take a great pride in the railway and all they have achieved. New engines will be running as time passes, giving added interest to those who enjoy miniature railways.

The last weekend in April is designated an open weekend when other enthusiasts bring their steam, electric and petrol-engined locos.

The fully-automatic signalling system really comes into its own at these times when large numbers of trains need to be carefully controlled to ensure safety. We also hold other special steam open days which are always great fun both for the enthusiasts and the general public.

Our woodlands took a terrible hammering in that terrifying hurricane in January 1990, as I wrote earlier. We hardly knew what had hit us. We lost more than 100 of our beautiful trees. At least 20 trees came down in our drive alone. A good deal of the foxproof fence surrounding the lake was broken by falling trees. Our noble staff spent hours patching everything up so that our precious birds would not escape - or, even worse, foxes get in and kill them.

We had to abandon the Tree Trail after that but much re-planting has gone on since then and the woodlands now look almost as lovely as ever.

Gardening gives me the greatest pleasure and has made me many friends, particularly through my contacts with the Rode Horticultural Society and the British Clematis Society.

I was very honoured recently when the British Clematis Society elected me as an Honorary Member when I decided to retire from my official duties with them. I could not have been in better company. Other Honorary Members elected years earlier were none other than Christopher Lloyd and Jim Fisk, two of our most famous and admired gardeners, the clematis author Dr John Howells and, more recently, Vince and Sylvia Denny who have sorted and packed thousands of packets of seeds.

I received a beautiful glass goblet from the society, hand engraved with drawings of the clematis Betty Risdon, named in my honour, and the most wonderful testimonial to my work.

CHAPTER TWELVE

Bird Thefts

I am afraid we have lost quite a few birds to thieves and I find this one of the most difficult, painful subjects to write about. There are many emotions - but anger is the dominant one.

Greed and money are the only motives of the people who steal birds. They have no feeling for the birds themselves who must suffer dreadful traumas after being caught and stuffed into sacks.

Nor is there any consideration for the owners and keepers who have devoted so much effort into looking after the birds and ensuring their well-being. They may have spent many years building up a breeding stock with loving care, only to find empty cages one day. Some of the birds which are stolen - often to order - are on the endangered list or part of a breeding pair.

In our early days we did have the odd bird stolen but with the help of the police and the media we always managed to get them back. I well remember having to go and collect an Amazon Grey Parrot from a police cell.

We have many great characters among our birds and the theft of a much-loved friend is particularly distressing. Mr and Mrs Honey were two favourites, our very tame pair of Moluccan Cockatoos who loved to be handled and who sometimes flew at liberty. One day Mr Honey went missing. As you can imagine, we were terribly worried.

I contacted the Press, TV and radio and happened to say jokingly during one interview that Mr Honey was quite partial to a gin and tonic. He had been on my arm one day when I was having a drink with some friends and turned away to speak . . . when I turned back, there he was with his beak in my glass. Don was furious with me for making the remark. He was afraid that Mr Honey would be given a lot of booze by whoever had found him.

It turned out that two boys had come to the gardens and decided to steal Mr Honey. The ring-leader offered another boy 2s 6d (12 1/2 p in today's money) but had to raise it 5s (25p) before the culprit agreed to do the deed and carry the cockatoo away in a bag. The boys were later found in a nearby village and sent to a remand home for a month while Mr Honey was restored to his rightful home.

In the mid-1970s we had a particularly bad theft when 10 African Greys were stolen from two of our aviaries. It was a horrible experience.

This was followed by the worst incident of all. It happened during Easter in 1993 when 19 birds worth more than £20,000 were taken. We had been pretty lucky until then because these sort of crimes were starting to happen all over the country. The criminals know exactly what they are after and usually have a buyer ready and waiting before they attempt the theft. Many of our friends in the bird world have suffered and we have had to increase our security to protect ourselves against thieves.

These thefts were heartbreaking to Don and to everyone else at Rode. We were always left wondering how many actually managed to survive their ordeal.

One whose fate we did learn about was our old friend Sigee, the Bare-fronted Sulphur Crested Cockatoo we had been given. The loss of Sigee depressed us for months and we never forgot him.

Then, years later, I spotted Sigee on a Vets School TV programme in 1996. I saw it was Sigeee at once. The lady who took him to Vets School in an attempt to save him said she was told he was 10 years old when she bought him. I have photographs of him taken in 1963 soon after he was presented to us when he had already spent the last 12 years in a pub.

Sadly, Sigee died soon after he appeared on the TV programme. The matter, as I write, is still in the hands of the police who are trying to trace the various people who handled Sigee after he vanished from Rode.

We have had several further thefts, part of a national problem which has seen £2-3 million-worth of birds stolen. One particularly clumsy incident here involved incredible damage to our aviaries by the thieves. It was followed just a week later by the professional theft of our much-prized pair of Hyacinthine Macaws. We live in hope that we will get them back one day. Certainly, identification of stolen birds is becoming much easier these days and we micro-chip our birds as a safety measure.

One bird garden near Weston-super-Mare was targeted so many times that it was forced to close. What a dreadful state of affairs.

Not so long ago we had yet another break-in. We were lucky this time. The man or men concerned were disturbed and had only managed to take four birds. We later discovered a pile of 23 sacks . . . they were clearly aiming for a very big clear-out.

There are two questions we are often asked when the subject comes up: "Can't you insure the birds? Do the police help?"

The answer to the former is that the cost is prohibitive. The answer to the latter is a little less straightforward.

Since the moment we opened our gardens, we have had excellent support from the grass-roots of the police force - constables and sergeants, particularly our local officer. No-one could have given us more help.

Unfortunately the higher ranks seem less keen to co-operate. Their attitude seems to be: "It's only a bird or birds."

We hear from time to time of Wildlife Liaison Officers' meetings in Somerset but we have never once had a visit from our area Wildlife Officer. There have been regular contacts with Wildlife Officers from the Thames Valley, Cheshire and Merseyside police forces and we liaise with them and exchange information about bird thefts. It seems very sad that this kind of thing is not done by our local force who pride themselves when the crime figures fall.

You could call this a plea from the heart: I know that if £20,000-worth of cash or goods was stolen from a building society, bank or jeweller's, action would be taken immediately. Yet thefts from bird centres are far worse. These are living creatures in need of expert care and special treatment. I recently checked a very large Stolen Bird File and the amount of missing livestock was quite unbelievable.

I believe that a useful move on the part of the police would be to make the main charge one of cruelty to animals, because that is precisely what it is.

All too often birds are taken at dead of night from homes where they are content, happy and well cared for. Shoved into sacks with other birds, they are frightened and will strike out and bite a friend, a mate or an offspring while being transported in the back of a van. It is impossible to imagine the physical or emotional state they will be in by the time they arrive at the thief's or thieves' destination. The effects on parrots, the most intelligent of birds, must be particularly distressing.

One gang of bird thieves was recently apprehended - and what happened? The longest sentence meted was two years imprisonment for one of the thieves and a derisory 12 months for the driver of the getaway vehicle. We have since discovered that this particular gang have understudies who are continuing their evil work. Clearly they do not see these sorts of sentences as deterrents.

One aspect of bird thefts which is rarely considered is the effect it has on staff who must spend sleepless nights worrying in case thieves strike again. And, too, our security today involves a great deal of extra work. Boundary fences need to be checked at all hours of the night and there is always that lurking doubt that thieves might be about.

Our staff are trained to watch out for any suspicious characters or suspect cars or vans. I am afraid that some visitors who are completely innocent are often under suspicion. I hope you now understand why.

David Nevill built up a dossier of bird theft information covering a number of years before he retired from the police. He was a great help and Len Simmons, from Linton Zoo, assisted him. John Hayward took over David's work and now co-ordinates the theft list with British zoos.

We have had a great deal of assistance from several contacts in the House of Lords and from several local MPs. Mike Curzon has visited some of the police forces with John Hayward. I can only hope that the new government will treat any future thefts more seriously.

CHAPTER THIRTEEN

The End of the Road

I hope this book has given you some idea of how we built up our very unusual business and all that it entailed and, too, that you have enjoyed reading about some of our travels.

Creating the bird garden has been hard work but it has always been exciting, full of challenges and often very amusing. Don had achieved his life-long ambition . . . running a zoo surrounded by the birds he loved so passionately and cared for so well.

Right at the start of this book I touched on what happened in the autumn of 1988 when Don began to have trouble with his eyes. By May 1989 Macular Degeneration had been diagnosed. There is still no known cure for the condition.

Life suddenly changed from that moment. Neither of us had any idea what it was or what it meant for our future. Macular Degeneration is the opposite of tunnel vision. There is sight around the periphery of the eye so the sufferer tends to look sideways at you. The magazine of the Macular Degeneration Society is called "Side View."

It took us ages to find out what to do and who to turn to for help. Eventually Social Services came to our aid and, too, we found several people living nearby who were suffering from the same condition.

One of the first things I was able to get help with was a closed circuit television system for Don through a business friend. Another local friend obtained Talking Newspapers. Later Don received Talking Books from the RNIB and various gadgets to help him live as normal a life as he could. It took a long time for him to accept carrying a white stick, but he did agree at last and it proved to be a great help.

The anger and frustration of suffering this disability never left Don in the final five-and-a-half years of his life. He developed deafness, which made matters even worse and in the last year of his life I had to have home nursing for several months. He was forced to spend a lot of time in a wheelchair, which he hated. Our doctor suggested that he should go into a home for respite care. He spent five months there and suffered a series of mini-strokes and, eventually, pneumonia.

I learnt so much during those last years of his life about facilities for the disabled. I recall booking a room in a hotel for a week. I told the receptionist that my husband could not see and she replied rudely: "What is that to me."

My complaints about this insensitive remark went up to the English Tourist Board and then on to the people handling the "Tourism For All" campaign. These experiences made me acutely aware of providing the best possible facilities for the disabled at Rode.

It was heartbreaking to watch Don's suffering but, with the help of my family and friends, I managed to cope.

When Don died, we held a small funeral service with family and friends at the little local church followed by a cremation. It was a lovely afternoon and, afterwards, we had tea outside the bungalow. One of our tourist attraction friends had also died and his funeral was held that day too. While we were having tea, a Hercules plane kept flying backwards and forwards over us. We cheered ourselves up by imagining that it was Don and his friend taking turns to drive.

The Memorial Service a month later was held in the large church in the village. A friend read "Iona Benedicite", our friend Geoffrey Greed from Bristol gave a reading and Johnny Morris gave a lovely address which had everyone laughing. Don loved fairground organs so we had one play-

ing at the bird gardens car park before we left for the ceremony and when we came back.

Our rector helped me choose the hymns and the last one
was to the tune of "I do like to be beside the seaside" so
everyone came out of the church still laughing. That was
just what Don would have wanted.

My thoughts were: Be happy for him that his suffering is
over. But I was sad for myself. I think of him on his cloud
with all his Zoo and birdy friends . . . organising a
"Heavenly Zoo and Bird Gardens".

So, Don, in your heavenly zoo, I dedicate this book to
you.

Betty Risdon